Quickstart
Molecular
Biology

An Introduction for Mathematicians,
Physicists, and Computational Scientists

ALSO FROM COLD SPRING HARBOR LABORATORY PRESS

A Genetic Switch, Third Edition, Phage Lambda Revisited

Experimental Design for Biologists

Genes and Signals (Kindle only)

Mammalian Development: Networks, Switches, and Morphogenetic Processes

RNA Worlds: From Life's Origins to Diversity in Gene Regulation

Signal Transduction: Principles, Pathways, and Processes

Quickstart Molecular Biology

An Introduction for Mathematicians,
Physicists, and Computational Scientists

Philip N. Benfey
Duke University and the
Howard Hughes Medical Institute

COLD SPRING HARBOR LABORATORY PRESS
Cold Spring Harbor, New York • www.cshlpress.org

Quickstart Molecular Biology
An Introduction for Mathematicians, Physicists, and Computational Scientists

© 2014 by Cold Spring Harbor Laboratory Press, Cold Spring Harbor, New York
Printed in the United States of America

Publisher	John Inglis
Acquisition Editor	Richard Sever
Director of Editorial Development	Jan Argentine
Developmental Editor	David H. Hatton
Project Manager	Inez Sialiano
Permissions Coordinator	Carol Brown
Production Editor	Kathleen Bubbeo
Production Manager	Denise Weiss
Cover Designer	Pete Jeffs

Library of Congress Cataloging-in-Publication Data

Benfey, Philip N.
    Quickstart molecular biology: an introductory course for mathematicians, physicists, and computational scientists / Philip N. Benfey, Duke University.
        pages cm
    Summary: "This book is an introductory course in molecular biology for mathematicians, physicists, and engineers. It covers the basic features of DNA, proteins, and cells but in the context of recent technological advances, such as next-generation sequencing and high-throughput screens, and their applications. This enables readers to move rapidly from the basics to an understanding of cutting-edge research in systems biology and genomics"--Provided by publisher.
    Includes bibliographical references and index.
    ISBN 978-1-62182-033-8 (hardback) -- ISBN 978-1-62182-034-5 (paperback)
    1. Molecular biology--Textbooks.   I. Title.
    QH506.B456 2014
    572.8--dc23
                            2014007639

10   9   8   7   6   5   4

All World Wide Web addresses are accurate to the best of our knowledge at the time of printing.

For a complete catalog of all Cold Spring Harbor Laboratory Press publications, visit www.cshlpress.org.

Contents

Acknowledgments

THERE ARE MANY PEOPLE TO thank when a book of this nature makes it to publication. Gary Carlson, former editor at Prentice Hall, commissioned the book and then was remarkably patient as it arrived in dribs and drabs. He provided insightful editing and guidance, and, when he announced his upcoming retirement, I realized I better finish it up. The editorial staff at Cold Spring Harbor Laboratory Press have been highly professional and efficient. Courtney Babbitt and Steve Haase contributed portions of the chapters on human evolution and biological oscillators, respectively, and Heidi Cederholm wrote a first draft of the glossary. The students in my class at Duke used earlier versions of this work as a textbook and provided helpful comments and feedback. During the six-plus years from conception to completion of this book, research in my laboratory was funded from a number of sources, including the National Institutes of Health (NIH), National Science Foundation (NSF), Defense Advanced Research Projects Agency (DARPA), Gordon and Betty Moore Foundation, and Howard Hughes Medical Institute. Finally, my deepest gratitude for the understanding and support of my wife, Elisabeth.

About the Author

PHILIP BENFEY IS the Paul Kramer Professor of Biology and an Investigator of the Howard Hughes Medical Institute at Duke University, Durham, North Carolina, USA. His research focuses on plant developmental genetics and genomics. He is a fellow of the American Association for the Advancement of Science and a member of the U.S. National Academy of Sciences. Dr. Benfey received his Ph.D. from Harvard University and a DEUG (Diplome d'Etudes Universitaire Generale) from the University of Paris.

CHAPTER 1

Introduction

1.1 Why This Book?

Shortly after I arrived at Duke, I met with an engineer who expressed an interest in biological networks. From that chance encounter developed a regular meeting of researchers who shared this interest. The "biological networks group" became the nucleus for a successful proposal to the National Institutes of Health to become a national Center for Systems Biology.

What characterized this group of researchers from the beginning was a nearly equal division among those trained as experimental biologists and those trained in one of the more quantitative scientific disciplines: mathematics, computer science, statistics, physics, and engineering. We realized that we would have to develop new means of educating each other across what is a fairly sizeable knowledge and cultural divide. It was not sufficient to tell our colleagues that they should read the standard textbooks in our respective fields. For the biologists, this would have meant using quantitative skills that had rarely been robustly acquired. For the theorists, it would have meant dealing with the standard biology textbooks, which tend to be off-putting in their size and level of detail. We experimented with various formats, and this book is the result of an approach that seemed to work for people who have backgrounds in the quantitative sciences but have had little or no formal training in biology. It has been used in a course at Duke University in which the students primarily major in electrical engineering, computer science, or mathematics.

1.2 How Is the Book Organized?

The goal of this book is to take the student quite rapidly from basic biological information to an understanding of cutting-edge technology and the results obtained from use of that technology. Each of the first four chapters is organized in a similar manner, beginning with the fundamentals of biological structure and how that structure is crucial for function. It then moves to a description of advanced-technology platforms used to analyze the biological entity. A final section addresses how data generated by each technology platform are analyzed.

The remaining chapters build on this knowledge base, with more in-depth exploration of examples of specific areas in which genomics and systems biology are having an impact. These include analysis of biological oscillators, the process of development from single cells to multicellular organisms, complex genetic traits, and insights into how the human genome is evolving. In a final chapter, I muse on the problem of using mathematics to analyze biological problems and look forward to a new era of quantitative "mathematical biology."

C H A P T E R   2

DNA: How Cells Store Information

2.1 What Does the Structure of DNA Say about Its Function?

The discovery of the structure of deoxyribonucleic acid (**DNA** [all terms in **bold** are defined in the Glossary]) is one of the most publicized detective stories in science. One of the key clues was that the building blocks, known as bases, came in four "flavors" and that there were approximately equal numbers of two sets of them. Another clue that happened to be observed by Jim Watson in Rosalind Franklin's laboratory was that an X-ray image of DNA revealed remarkable symmetries. Yet, in a race between Nobel laureate Linus Pauling and the young upstarts Watson and Crick, it took weeks of trial-and-error assembling cardboard cutouts to come up with a structure that made sense.

The specific pairing between bases that Watson and Crick discovered is both the key to how genetic material is replicated and provides a means of copying the information stored in DNA and translating that information into proteins—the primary workhorses in a cell. This is what became known as the **central dogma**. In its initial form, the dogma was that information cannot flow backward from protein to DNA. As more was understood about this process, it morphed into its current form: Information is transferred from DNA into RNA (a related molecule) and then into protein. Genes are the units of DNA that get "translated" into individual proteins. What has become clear in the past 20 years is that DNA does much more than provide the information for production of proteins. The sequencing of the entire genomes of many organisms has led to the identification of a host of new regulatory molecules encoded by DNA as well as the realization that genomes are in dynamic flux.

3

2.2 What Is DNA?

DNA is a very long molecule that contains the information used by cells to function and reproduce. As everyone knows, the overall structure is a double helix. But this aspect of its structure does not reveal much about how it encodes information or how this information is copied during cell division.

Of crucial importance to the function of DNA is the nature and structure of its constituent **bases**. The four bases, A, T, C, and G (**adenine, thymine, cytosine, and guanine**), all form relatively flat chemical structures that have reactive components along their outer edges (Fig. 2.1). Each base is chemically linked to a sugar molecule (deoxyribose), and the combination of base and sugar molecule is called a **nucleotide**. The shape and placement of reactive components of the bases result in the base A being able to fit nicely opposite the base T, and the base C fitting nicely opposite the base G (Fig. 2.1). In DNA, the nucleotides are linked to each other through a molecular bridge that contains the atom phosphorous (Fig. 2.2). The phosphorous atoms in DNA have a strong negative charge, which tends to push the two **phosphate backbones** of the DNA molecule away from each other, just as magnets with similar polarity force each other apart. The combination of the flat, complementary bases and their phosphate backbones results in the helical structure of DNA (Fig. 2.3). The chemical structure of the individual DNA strands also provides polarity to each of the strands. The carbon atoms in the sugar are numbered from 1 to 5, and the phosphate is attached to the fifth carbon (C5) on one nucleotide and to the third carbon (C3) on the adjacent nucleotide (Fig. 2.2). A short stretch of nucleotides is

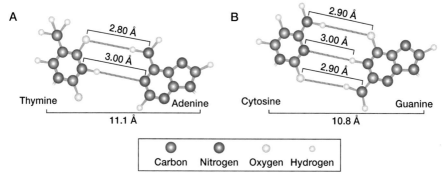

Figure 2.1. The bases that comprise DNA have a flat chemical structure and can be aligned so that chemical bonds form between (A) thymine and adenine (T, A) or (B) cytosine and guanine (C, G). Note that there are three bonds between C and G and only two between T and A. This results in a stronger interaction between C and G than between T and A.

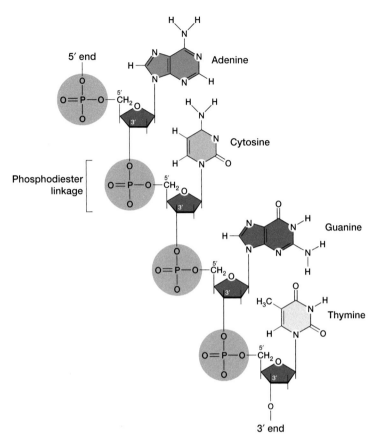

Figure 2.2. A strand of DNA comprises nucleotides (A, T, C, G) that are linked together through phosphate molecules (PO_4), which form phosphodiester linkages. There is a polarity to the DNA strand, which is determined by the phosphate linkage being attached to the 5' carbon (C) of one nucleotide and the 3' carbon of the next nucleotide.

called an **oligonucleotide**, which has two ends, one end with a phosphate attached to the C5 carbon—the 5' (5 prime) end—and the other end with a phosphate attached to the C3 carbon—the 3' (3 prime) end (Figs. 2.2 and 2.3).

2.3 How Is Information Transferred during DNA Replication?

When a cell divides, it must faithfully reproduce its DNA so that each of the two daughter cells gets the same genetic information. This process is called **DNA replication**, which occurs in the nucleus of dividing animal and plant cells. What Watson and Crick realized was that the complementarity of the

A

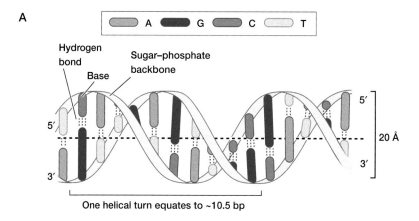

B

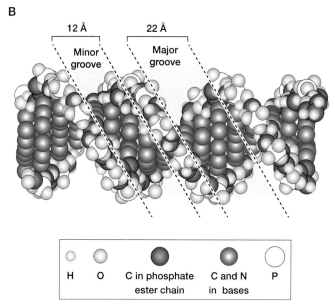

Figure 2.3. When two DNA strands with complementary bases anneal together, they form a double helix in which each helical turn comprises approximately 10.5 base pairs (bp) (A). The helix forms such that there is a more open part called the major groove and a less open region called the minor groove (B).

bases in a DNA molecule provides an elegant means of copying the information they contain. First, the double helix needs to be "unzipped" to expose the individual bases, and then two new molecules are made by adding complementary bases to the template of the two exposed strands (Fig. 2.4). This sounds straightforward, but in fact it requires the help of a remarkably well-adapted **enzyme** (a protein capable of facilitating chemical reactions) produced by the cell called **DNA polymerase** and a set of ancillary molecules.

The job of DNA polymerase is to recognize DNA when it is unzipped, and thus **single-stranded**, and to add the appropriate complementary base to a growing strand. To do this, DNA polymerase requires that there be an adequate supply of the four nucleotides A, T, C, and G in the cell so that it can add them as needed. The nucleotides are always added in one direction, from the 5'

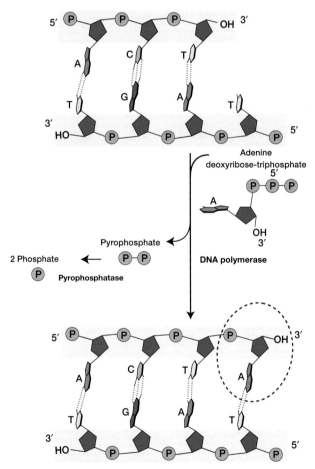

Figure 2.4. Replication of DNA occurs by addition of a complementary base (in this case, an A opposite an unpaired T) by DNA polymerase. The energy for the reaction is derived from removal of two phosphates (pyrophosphate) from the incoming nucleotide (adenine deoxyribose-triphosphate) by the enzyme pyrophosphatase.

direction to the 3′ direction (Fig. 2.4). The addition of the bases and the movement of DNA polymerase along the single-stranded DNA template requires energy, which it obtains by removing two of the three phosphate molecules found on each free nucleotide before insertion into the growing DNA strand (Fig. 2.5). The removal of the two phosphates breaks high-energy chemical bonds, and this is the energy that drives the process of replication.

Even with the discovery of the structure of DNA, it was not at all obvious how replication would proceed. For example, it could be that both strands were copied, but the template strands immediately came back together (a process called **annealing**) to restore the original "parental helix" and make a new double helix. A second possibility was that the strands were copied and then annealed to the

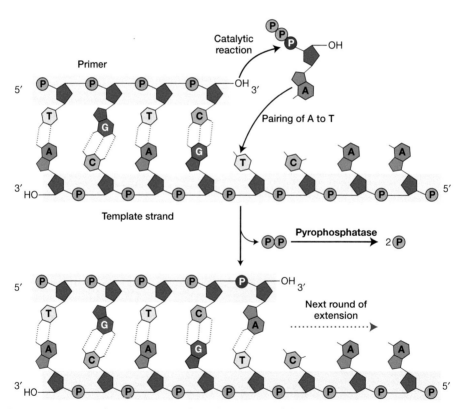

Figure 2.5. DNA polymerase can only replicate DNA if there is already a double-stranded region present. During replication, the double-stranded region is formed by the addition of primers. In this example, the incoming A nucleotide forms a phosphate linkage to the free 3′ end of the C in the primer and base-pairs with the complementary T base in the opposite strand, thus extending the chain by one nucleotide.

original templating strands, whereas a third possibility was that there was random copying of one strand or the other. In a famous experiment by Matt Meselson and Frank Stahl, bacteria were fed heavy nitrogen (N^{15}), which would be incorporated into the DNA, for the time it takes for one round of replication, then fed normal nitrogen (N^{14}) after the first round of replication. DNA was isolated from the bacteria before feeding the N^{15}, after one round of replication on N^{15}, and after one round on N^{15} and two rounds of replication on N^{14}. The DNA from the different samples was put into tubes with a gradient of cesium chloride and spun in a high-speed centrifuge, causing the DNA of different densities to form distinct bands in the centrifuge tube. Meselson and Stahl observed that, when the bacteria were grown continuously on N^{15}, their DNA ended up at the bottom of the tube, indicating it had a high density and that all the DNA had taken up the heavy nitrogen. After one round of replication on N^{15}, the DNA was found in the middle of the tube, and after two subsequent rounds of replication on N^{14}, some of the DNA was still in the middle, whereas some was at the top of the tube. Meselson and Stahl's interpretation was that DNA was replicated in a **semiconservative** manner, meaning that, as new strands were formed, they base-paired with the existing strand (Fig. 2.6). If replication were conservative, then the investigators would not have gotten the band in the middle of the tube. If it were random, then they would not have gotten two bands after two rounds of replication.

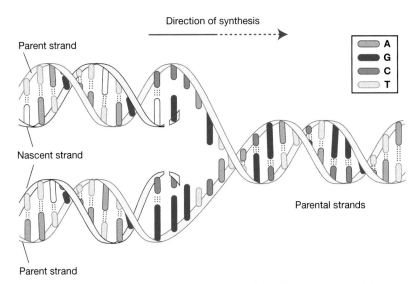

Figure 2.6. Semiconservative DNA replication occurs through the copying of the two parental DNA strands and the formation of two new double helices, each containing one of the original parental strands and one nascent strand.

Perhaps surprisingly, DNA polymerase cannot replicate DNA unless it has a piece of DNA or RNA already in place. RNA is a molecule that resembles DNA but has a slightly different chemical structure and is called RNA as it is a **ribonucleic acid**. In other words, DNA polymerase cannot just sit down on a single-stranded piece of DNA and start adding bases. It has to have a **primer**, a piece of DNA or RNA that is annealed to the single strand, to make at least a short part of it double stranded (Fig. 2.7). The need for a primer presents a conundrum for DNA polymerase during replication. What can it use as a primer to initiate replication once the DNA helix is unwound? It turns out that it uses a small piece of RNA, which is

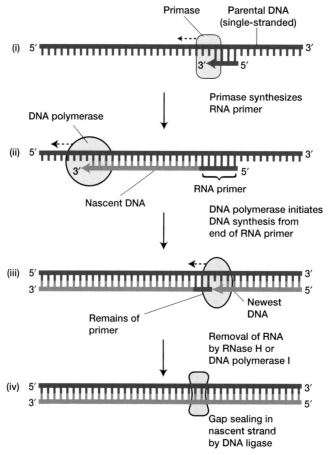

Figure 2.7. Lagging-strand DNA replication occurs through the use of a primase (i) to make an RNA primer, replication by a DNA polymerase (ii), removal of this primer after DNA synthesis by the RNase H enzyme together with filling of the gap by DNA polymerase (iii), and finally ligation by DNA ligase (iv).

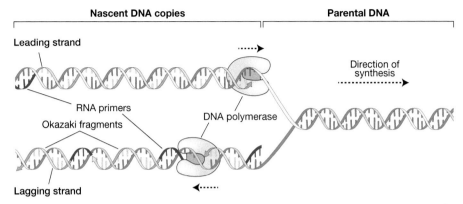

Figure 2.8. Leading- and lagging-strand replication use different mechanisms to complete the copying of the parental DNA strands, grasped by the hand-like polymerase structure.

copied from the DNA sequence by a special enzyme called **DNA primase** (Fig. 2.7). DNA polymerase then extends the short piece of RNA, adding DNA nucleotides, one at a time. When the structure of DNA polymerase was determined, it seemed to resemble a "hand" that grasped the template and primer and added the bases in its "palm" (Fig. 2.8). The incoming bases arrived through an opening between the "thumb" and curled "fingers" (Fig. 2.8).

Another problem presents itself to the polymerase through the fact that it can only move in the 5' to 3' direction. Once the DNA helix is opened and an RNA primer is made, DNA polymerase can move down the strand that has its 5' end at the opening, without a problem. This is called the **leading strand**. But what can it do with the other strand that has its 3' end at the opening? The solution to replicating the **lagging strand** is somewhat baroque. As the DNA helix is opened, primase makes an RNA primer at the opening, then DNA polymerase extends the primer away from the opening. This has to be repeated for the lagging strand over and over again as the opening or **replication fork** moves along, leading to the formation of a patchwork of RNA primers and short segments of DNA (Fig. 2.7). These short segments are known as **Okazaki fragments**. The problem then is to get rid of the RNA primers and make a continuous DNA molecule from the lagging strand. An enzyme called **RNase H** chews away the RNA primer (Fig. 2.7). DNA polymerase can then use the piece of DNA that is 5' of where the RNA primer sat as a primer and can extend up to the next piece of DNA (Fig. 2.7). There remains one final problem—how to join the two pieces of DNA. This is performed by yet another enzyme, called **DNA ligase** (Fig. 2.7). Once all of these operations are performed, the lagging strand is then a single contiguous piece of DNA.

2.4 How Is Information Stored in DNA?

In addition to providing a remarkably elegant means for reproducing the information in DNA, the four bases A, T, C, and G are also used in a three-letter code to spell out **amino acid** sequences for making proteins (Fig. 2.9). There are 20 different amino acids used to make proteins, but there are 64 (4^3) possible three-letter combinations, called **codons**, that can be made from the four bases. For scientists to determine which codons specify each of the 20 amino acids, synthetic molecules of various combinations of the four letters were used to prime protein synthesis *in vitro* (in a test tube). It turns out that the potential codons are unevenly distributed among the 20 amino acids. Some amino acids, such as methionine, are specified by only a single codon, ATG. For other amino acids

Position #2

		U		C		A		G		
Position #1	**U**	U U U	Phe	U C U		U A U	Tyr	U G U	Cys	U
		U U C	Phe	U C C	Ser	U A C	Tyr	U G C	Cys	C
		U U A	Leu	U C A		U A A	Stop	U G A	Stop	A
		U U G	Leu	U C G		U A G	Stop	U G G	Trp	G
	C	C U U		C C U		C A U	His	C G U		U
		C U C	Leu	C C C	Pro	C A C	His	C G C	Arg	C
		C U A		C C A		C A A	Gln	C G A		A
		C U G		C C G		C A G	Gln	C G G		G
	A	A U U	Ile	A C U		A A U	Asn	A G U	Ser	U
		A U C	Ile	A C C	Thr	A A C	Asn	A G C	Ser	C
		A U A	Ile	A C A		A A A	Lys	A G A	Arg	A
		A U G	Met	A C G		A A G	Lys	A G G	Arg	G
	G	G U U		G C U		G A U	Asp	G G U		U
		G U C	Val	G C C	Ala	G A C	Asp	G G C	Gly	C
		G U A		G C A		G A A	Glu	G G A		A
		G U G		G C G		G A G	Glu	G G G		G

Position #3

Figure 2.9. The genetic code comprises the different three-nucleotide "words" that code within the mRNA for specific amino acids, which are incorporated into proteins during the process of translation. The three "stop" codons signal termination of production of the protein at the end of each round of translation of the mRNA.

such as serine, there are as many as six different codons, whereas most of the amino acids have between three and four codons each (Fig. 2.9). There are also three so-called stop codons, whose function is to signal termination of production of the protein. There is no start codon, but most proteins begin with the amino acid methionine.

The intermediate for protein production is a special type of RNA known as **messenger RNA or mRNA**. In a manner analogous to DNA replication, mRNA is copied, base by base, from the DNA template in a process called **transcription**, also occurring in the cell nucleus. The production and regulation of mRNA synthesis will be discussed in depth in Chapter 3. mRNAs exit the nucleus via specialized nuclear pores to carry the information into the cytoplasm, where it is ultimately translated into proteins.

There is information in DNA in addition to that used to specify the amino acids through use of codons. Of particular importance is information that tells the organism when and where to make mRNAs. Frequently, but not always, the DNA in front of the protein-coding region (5' of the start of transcription) provides these directions to the cell. It does so by encoding specific binding sites for a special class of proteins, known as **transcription factors**. After binding to the DNA, these proteins interact with each other and with the machinery that initiates production of RNA.

The DNA in an organism is organized into **chromosomes**. In bacteria, there is usually a single circular chromosome. In humans, there are 23 pairs of chromosomes, with some containing more DNA than others. All the DNA in all the chromosomes constitutes what is referred to as the **genome**. In humans, the **haploid** genome (constituting one member of each chromosome pair) is approximately 3.2 billion base-pairs (bp) and codes for approximately 23,000 genes.

2.5 What Is a Gene?

I have introduced the term **gene** without giving it a formal definition. This was intentional because there is no definition that is universally agreed upon. Originally, the term came from the field of **genetics**—the study of how mutations affect the observable attributes of an organism. For example, the "white" mutation in the fruit fly causes the fly's eyes to change from their normal red to white. Decades before the structure of DNA was determined, the fundamental properties of genes were worked out by geneticists. It was deduced that genes were particulate in nature—that is, they conferred one attribute or another (e.g., white or red). One form of a gene did not confer a range of colors. Genes were at specific positions on chromosomes, and the distances between genes could be determined by how often different combinations were found in the progeny when two parents with different **alleles**—genes with slightly different sequences—were mated.

When the structure of DNA became known and the genetic code was determined, the definition of a gene seemed fairly straightforward: It was the portion of DNA (generally ~2000 bp) that encodes a protein plus the portions upstream (5') and downstream that are found in the mRNA or are involved in its regulation. Complications arose in the 1970s when it was discovered that many protein-coding regions were interrupted by DNA sequences, which were termed **introns**. The protein-coding portions on either side of an intron were called **exons** (Fig. 2.10). The cell would remove introns using a process called **splicing**. Then it was found that different proteins could be produced by a cell by selectively using certain exons within a gene and not others. This is called **alternative splicing** as different introns and exons are spliced out to give rise to the final mRNA. So, does each one of the alternatively spliced messages represent a gene, or is the combination of all of them the gene? Most biologists now view the combination of possible alternatively spliced mRNAs as coming from a single gene. This is why it is said that the human genome has around 23,000 genes and encodes perhaps as many as 50,000 proteins. In addition, genes are no longer restricted to the portions of DNA that code for proteins. One example is ribosomal RNA genes, which produce a long RNA that serves a structural role; another is **microRNA** molecules, which regulate the activity of mRNAs. As we come to understand more about the various functions of the DNA in a genome, it is likely that the repertoire of meanings encompassed in the term "gene" will only increase.

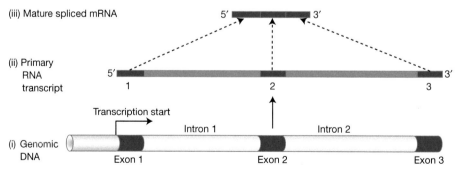

Figure 2.10. In eukaryotes, the information in DNA that will be translated into proteins is interrupted by regions called introns (i). The introns are present in the primary RNA transcript (ii) and are subsequently spliced out (iii), leaving the mature messenger RNA (mRNA), which, upon export from the nucleus, is the substrate that is translated into protein.

2.6 How Is the Sequence of DNA Determined?

Arguably, the biggest breakthrough in biology since the discovery of the structure of DNA is the ability to determine the entire sequence of the DNA in an organism. **Genome sequencing** has made the entire repertoire of information stored in the DNA available for analysis. In addition to providing a nearly complete "parts list" of all the genes in certain organisms, comparison between genomes has revealed what is likely to be functional in the genome and what is random noise. The list of organisms whose genome has been sequenced becomes longer every day, and the rate of sequence acquisition is increasing exponentially. What were the technical developments that allowed entire genomes to be sequenced in a cost-effective manner, and what is driving the latest advances in genome sequencing technology?

In the 1970s, two means of determining the sequence of DNA were developed. Both relied on advances in **gel electrophoresis** to separate fragments of DNA according to their molecular weight. In gel electrophoresis, molecules are subjected to an electric field that causes them to move through a polymerized material. Two common polymers used to make the gel are agarose and acrylamide. The former has a consistency similar to gelatin, whereas the latter tends to be stiffer. The molecules to be separated are loaded into wells at one end of the gel, and a solution containing electrolytes is added. A voltage differential is then applied, causing the electrolyte solution to move through the gel. The molecules in the wells are also subjected to the voltage differential and will begin to move in the direction determined by their charge (Fig. 2.11). Owing to the nature of movement through the gel, the rate of movement of molecules is inversely proportional to the square of their molecular weight. Thus, smaller molecules will move much faster than larger molecules (Fig. 2.11).

For separation using gel electrophoresis, DNA has the fortunate properties of being electrically charged and having the charges relatively evenly distributed along the length of the molecule. The negative charges on DNA come from the phosphates in its backbone (Fig. 2.2). Another fortunate property is that DNA is a linear molecule that can be straightened out fairly easily. Combined, these attributes mean that DNA molecules separated by gel electrophoresis will be ordered almost perfectly by length. That is, when the current is flowing from top to bottom, the longest molecules will end up toward the top of the gel, and the shorter molecules toward the bottom (Fig. 2.11). The key breakthrough in sequencing was to use ultrathin acrylamide gels that permitted a resolution of one base difference between the DNA molecules separated by the gel. The original automated sequencers worked by the same principles—but used thin capillary tubes filled with polymers, as opposed to large slab gels, to separate the DNA.

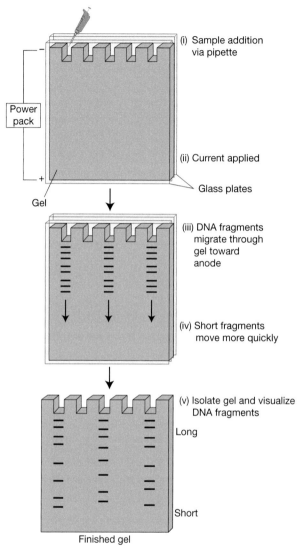

Figure 2.11. Gel electrophoresis is used to separate a mixture of DNA fragments according to their molecular weight. The DNA fragments are loaded into a well in the gel (i), then a current is applied (ii) and the negatively charged DNA migrates toward the positive pole (iii). Shorter DNA fragments migrate faster than longer fragments (iv) and thus end up nearer the bottom of the gel, as visualized by the ladder of fragments in the finished gel (v).

Once able to separate DNA by length, the next challenge was to produce ordered fragments of DNA, each one terminating with a known base. Walter Gilbert and Fred Sanger shared a Nobel Prize in 1980 for developing two different methods to do this. Gilbert's method, which is rarely used today, employed combinations of specialty chemicals to cleave differentially at either A, T, C, or G. Sanger's approach, by contrast, has proved remarkably robust and, until recently, was the methodology underlying most DNA sequencing. What Sanger did was harness DNA polymerase to generate strands of DNA forming a "ladder" of all fragments that end with each of the four bases. The key to the process was a particular chemical modification of each of the bases so that DNA polymerase could add it to a growing strand, but could not extend the polymerization reaction by adding an additional base to it. These modified bases are called **dideoxynucleotides** and they cause the chains of DNA to terminate, hence the name of the approach: **chain-termination sequencing**. The question then was how to get DNA polymerase to form a ladder of all possible lengths of polymerized DNA fragments, rather than stopping at the first occurrence of the base complementary to the dideoxynucleotide. The trick was to mix the right proportion of normal nucleotides with the dideoxynucleotides so that there was a relatively low probability that DNA polymerase would insert the dideoxynucleotide. For example, if the right proportion of dideoxy A is added to normal A, then the newly made DNA strand will terminate at positions corresponding to each location that there is a T in the template (as A pairs with T). To be able to read the entire sequence of a region of DNA requires a separate reaction for each of the four bases: A, T, C, and G. Because DNA polymerase cannot work without a primer, an oligonucleotide that is complementary to the first 18–24 bases to be sequenced has to be added as well. To simplify matters, most sequencing reactions are performed such that the first stretch of bases is complementary to a standard **sequencing primer**.

Once the sequencing reaction has been run, the molecules generated by the reaction have to be detected. This was originally achieved by making the chains radioactive by using radioactively labeled nucleotides in the reaction. For the first generation of automated sequencers, fluorescent labels were substituted for radioactivity. A different colored label was used for each of the four sequencing reactions (A, T, C, and G), which allowed them to be combined in a single capillary tube (Fig. 2.12).

2.7 How Are Genomes Sequenced and Assembled?

The federally funded effort to sequence the human genome cost ~$3 billion dollars and took more than 10 years. Within the near future, the cost is likely to be less than $1000 and take less than a week. How has this dramatic change in sequencing come about?

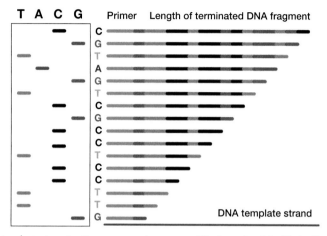

Figure 2.12. In chain termination sequencing, a modified nucleotide (T, A, C, or G) is incorporated in the DNA strand during replication, which prevents further polymerization. A separate reaction is run for each nucleotide, resulting in DNA strands whose lengths are indicative of the position of each of the nucleotides (e.g., each occurrence of thymine [T]) implicated in that reaction.

The initial draft of the human genome was performed by Sanger sequencing and involved a series of discrete steps. The first was to make a collection known as a **library** of large fragments of genomic DNA. The second step, called **production sequencing**, involved generating smaller DNA fragments to be sequenced, performing the actual sequencing reactions, and running the reactions through an automated sequencer. The third stage was called finishing—this involved assembling the raw sequences (called **sequence reads**) into a continuous sequence and then filling any remaining gaps in the sequence. Gap filling was performed using the **polymerase chain reaction (PCR)** described below, with primers based on sequences found on either side of the gap.

One of the reasons that the publicly funded sequencing effort, known as the **Human Genome Project**, was so costly and took so long was that it used a **map-based strategy**. This required finding library fragments that overlapped from one end of a chromosome to the other. Each of these large fragments was broken into overlapping smaller fragments, which could be sequenced using chain-termination sequencing. Finally, the sequence of the small fragments was assembled by finding their overlaps, and this process was repeated for the large overlapping fragments.

An alternative approach was introduced by the biotechnology company Celera and its founder Craig Venter, called "shotgun sequencing." They challenged the publicly funded effort, claiming that, by using Celera's **whole-**

genome shotgun approach, they could sequence the genome for one-tenth the cost and in a much shorter time. With shotgun sequencing, the whole genome is broken into small pieces that can be readily sequenced. The challenge is to find the overlapping portions among the huge number of small sequences so that a contiguous sequence can be assembled. To do this required the development of robust computer programs that ran on the fastest computers available at the time. In the end, the publicly and privately funded efforts completed the human genome sequence at about the same time, with each claiming "victory."

2.8 How Are Sequencing Methods Changing?

When the Human Genome Project began in 1990, large sums of money were set aside to develop new and better sequencing methods. Remarkably, the human genome was completed without any major new advances in sequencing technology. However, over the past decade, several new technologies have been introduced that are rapidly changing the uses and economics of sequencing. Together, they are referred to as **next-generation sequencing**. The drop in the cost of sequencing has been striking, progressing at a much faster pace than even Moore's law, which has accurately predicted the reduction in cost of computing over the past few decades.

Most of the next-generation sequencing methods rely on the PCR to generate sequence reads. When it was introduced in the 1980s, PCR had a profound effect on the field of biology. The key to making PCR work was the discovery of DNA polymerases that would still function at near-boiling temperatures. These were initially isolated from bacteria that live in hot springs. As with all DNA polymerases, these thermostable polymerases require a template and a primer to generate DNA. In the polymerase chain reaction, two primers are used, one that is complementary to the 5′ end of the template, and the other that is complementary to the 3′ end. The reaction is performed in an instrument in which the reaction temperature can be rapidly and precisely raised or lowered.

For example, if the template is double-stranded DNA, the first reaction raises the temperature above 90°C, which causes the two strands to come apart (Fig. 2.13). Then the temperature is reduced so that the two primers can anneal, one to each of the two DNA strands. The temperature at which this annealing occurs depends upon the relative numbers of Gs and Cs to As and Ts in the primers. The greater the GC content, the higher the temperature. The reaction temperature is then raised again to ~70°C, and the thermostable polymerase begins

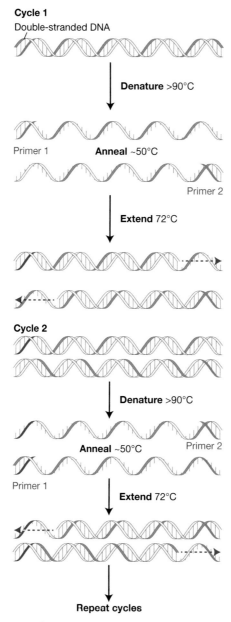

Figure 2.13. The polymerase chain reaction (PCR) employs two primers that anneal to each of the strands of DNA after it is denatured. A heat-stable DNA polymerase then extends complementary strands from each primer. Once the two double-stranded DNA molecules are formed, they are denatured, and the cycle repeated, resulting in an exponential increase in DNA concentration.

to generate DNA strands from each of the primers. When this synthesis is completed, there are now two newly formed double-stranded DNA molecules equivalent to the starting DNA template (Fig. 2.13). The process is then repeated so that, at each cycle, there is an exponential increase in DNA. The remarkable power of PCR is its ability to start with as little as a single DNA molecule and generate as much DNA as one could possibly use.

2.9 What Is Next-Generation Sequencing?

The first next-generation method to be widely used was based on a process called **pyrosequencing**. As with Sanger sequencing, pyrosequencing uses DNA polymerase to incorporate nucleotides complementary to a template. However, instead of using chain-terminating nucleotides and observing relative chain length after the reaction is completed, in pyrosequencing the incorporation of each nucleotide is observed in real-time through the release of **pyrophosphate (PPi)**. As each nucleotide is added to the growing DNA chain, one molecule of pyrophosphate is released. To be able to observe the amount released, an enzyme is added that converts pyrophosphate to the energy-rich molecule ATP. An enzyme that comes from fireflies, called luciferase, uses the ATP and another molecule, luciferin, and generates light. The precise amount of light produced in the reaction is directly proportional to the amount of pyrophosphate released in the sequencing reaction. Thus, to determine the DNA sequence, each nucleotide is added individually and the amount of light generated is measured. For example, if one G is added by the reaction, a single molecular equivalent of light is observed. However, if two Gs are added, then twice the amount of light would be seen (Fig. 2.14).

The use of pyrosequencing (sometimes referred to as **454 sequencing** after the company that initially commercialized it) for large-scale genome sequencing involves several steps. The first is random shearing of the genomic DNA. Then short pieces of double-stranded DNA called **adapters** are ligated onto both ends of the randomly sheared fragments. The DNA is then denatured and allowed to anneal to oligonucleotides complementary to the adapters, which are attached to special beads. This is done so that each bead is attached to a single genomic fragment. PCR is then performed in a water-in-oil emulsion, causing the DNA to be amplified several millionfold. The beads are then distributed into the wells of a specially designed plate that can hold more than a million beads, each in its own well. The pyrosequencing reaction is performed simultaneously in all the wells with the addition of one nucleotide at a time. Light generation is recorded by a highly sensitive camera, which tracks the amount of light emitted from each well as each nucleotide is added. This is then translated into sequence reads.

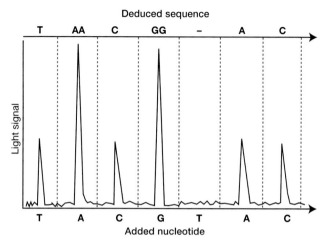

Figure 2.14. In pyrosequencing, a photon of light is emitted each time a nucleotide is added. The sequence is read by determining the amount of light emitted when each nucleotide is added to the reaction.

Currently, the most widely used next-generation approach is known as **Illumina sequencing**, and this is also dependent on DNA polymerase adding bases to a template. However, in Illumina sequencing, single DNA molecules are amplified by PCR after being fixed to specially coated slides. A crucial difference with Sanger sequencing is that chains are not terminated using dideoxynucleotides. Instead, all four bases are added at once, but each base has been chemically modified so that it can reversibly terminate the chain and emit fluorescent light. The different colors emitted at each point on a slide are detected by sensitive cameras. After detection, the bases are chemically modified so that they can accept another round of synthesis. By recording the different colors emitted at each spot over time, the Illumina sequencer generates a readout of the sequence for each genomic fragment.

These next-generation technologies have managed to reduce greatly the cost of sequencing while dramatically increasing throughput. One of the current drawbacks is that read-lengths tend to be relatively short (less than 500 bp). This is primarily a problem for de novo assembly of genomes that have never been sequenced before, but also can cause problems in resequencing genomes that contain many repetitive elements (discussed below).

A single-molecule sequencing platform that is currently on the market is made by Pacific Biosciences (PacBio). Their sequencer uses an engineered metal surface with wells 70 nm wide to trap individual DNA molecules.

Once constrained in the well, the DNA is unwound and primers are added so that both strands can be sequenced simultaneously. The major novelty is that there is no wash step required. Detection occurs as nucleotides are added in real-time. PacBio claims read-lengths of up to 10,000 bases at a rate of 10 bases per second. The key technical challenge is a high error rate in the sequence as it is read in real-time.

Among the newer approaches that promise longer reads as well as higher throughput, several are based on sequencing individual DNA molecules. One of these **single-molecule sequencing technologies** hopes to capture information from the different bases as a strand of DNA passes through a specially designed channel called a **nanopore**. Results indicate that each base has a slightly different charge that can be detected at the nanopore. The promise of this approach is that long DNA molecules could be sequenced in a matter of seconds to minutes.

2.10 How Are Sequence Data Analyzed?

No matter how DNA is sequenced, the sequence reads have to be assembled into contiguous segments. This problem is particularly acute for methods that generate short sequence reads. The starting point for any assembly is the comparison of two DNA sequences. With sufficiently long overlapping sequences that are identical, there is no problem except in the case of repetitive sequences (see below). However, no sequencing method is error free. Based on analysis of the errors contained in actual sequence reads, scoring systems have been devised to determine how likely it is that two sequence reads were derived from the same genomic region. For example, insertions or deletions of bases are scored differently from changes in the identity of the actual bases. The scoring system can then be used to identify high-scoring alignments that are likely to be overlaps. This is an active area of current bioinformatics research.

Once genomes are assembled, they are annotated, which involves finding all of their important features. The first step is usually identifying the regions that code for proteins. A search is made for sequences that could be stretches of codons uninterrupted by stop codons. Between these **open reading frames (ORFs)**, another search is made for sequences known to mark the borders between exons and introns and for sequences found immediately before and after coding regions. Although seemingly straightforward, current gene-prediction tools are imperfect. The most widely used tools depend on **machine learning algorithms** that find patterns based on a training set of sequences known to contain the desired features. Some of these tools are based on neural networks, whereas others search for **hidden Markov models (HMMs)**.

2.11 What Has Genome Sequencing Taught Us?

In addition to use in assembly, sequence alignments can be used to determine how the DNA sequence has changed during evolution. For example, comparison of the sequence of amino acids in a protein from humans and chimpanzees can be used to identify changes in the sequence that might have subtly altered its function. Alternatively, when a new gene is found that is involved in a disease, searching through all other sequenced DNA can point to related genes whose function is known.

Although there are several different means of performing sequence alignments, the most widely used is called **BLAST** (**basic local alignment search tool**). It is fast and effective but is based on a heuristic algorithm—thus cannot be proven to always identify the best alignment. Its approach is to break the sequence that one is using as a query into small overlapping pieces or "words." A search of the sequence database is then run with these words, and the highest-scoring hits are retained. Subsequently, the algorithm searches for an alignment that maximizes the similarity score between the query sequence and the DNA sequence found in the database. To determine how likely the alignment would occur by chance, one of the sequences is randomly scrambled multiple times, and the alignment score is recalculated. Comparing the results of repetitive random scrambling with the initial alignment score generates a probability that the alignment is not by chance.

A BLAST search can be run, free of charge, at the National Center for Biotechnology Information (NCBI) website (http://blast.ncbi.nlm.nih.gov/Blast.cgi). One can choose the type of alignment desired (e.g., DNA or protein) and the public sequence databases to be searched. The results are color coded to identify sequence alignments that are identical or different.

Once genome sequences started to become available, one of the first questions asked concerned how and why genome size varies among organisms. It seemed reasonable to assume that the larger the genome, the more genes there would be in it. Thus, a human genome with more than 3 billion bases should have far more genes that a fruit fly genome with around 120 million bases. In fact, the difference in gene number is less than twofold, raising the obvious question regarding what makes up the rest of the human genome.

Originally termed "junk DNA", the parts of the genome that do not code for proteins can be divided into different categories. There are sequences whose job is to make sure that the gene is only active in specific tissues or in response to specific stimuli. These regulatory sequences tend to be upstream of the coding sequences, although they can be separated from the gene they control by more than 50,000 bp. But the primary reason for the large differences in genome size is the number of **repetitive DNA elements**. As the name suggests, these are segments of DNA found throughout the genome that have very similar sequences. **Transposable elements** or **transposons** are DNA regions that are able to move between locations in the genome and make up a large part of the repetitive DNA. Transposons generally are several kilobases in length and have nearly identical sequences at each end. They can code for proteins, including a **transposase** that facilitates their movement. In some cases, they move by excising from one DNA location and inserting at another. A particular class of transposons called **retrotransposons** uses an RNA intermediate to make a new copy, while leaving behind the old copy. Thus, it is easy to see how these elements can rapidly increase their numbers in the genome—indeed, nearly 45% of the human genome is made up of transposons.

It has been shown in some plant species that there can be dramatic increases in the number of copies of certain retrotransposons within a few generations. The role of repetitive sequences, in general, and transposons, in particular, is still a matter of intense research. In the 1950s, Barbara McClintock studied the variegated color patterns found in some corn kernels. From the changes in the patterns of variegation, she developed a theory that predicted that genetic instability was important to evolutionary change. Later work showed that the variegation McClintock studied was due to the activity of transposable elements. There is increasing evidence that, in fact, transposition is induced under stress conditions, resulting in remodeling of the genome. In this way, large-scale changes can occur in the genome in relatively short periods of time, generating more genetic diversity upon which natural selection can work.

C H A P T E R   3

Gene Expression: How Cells Process Information

3.1 What Is RNA?

When you walk into a molecular biology laboratory you will not find most researchers wearing lab coats, but nearly everyone wears vinyl gloves. The gloves serve two purposes: to keep nasty chemicals off one's hands and to make sure that RNA does not get degraded. The chemical structure of RNA is very similar to that of DNA—the only difference being the presence of an additional oxygen and hydrogen atom at one position in the RNA molecule (Fig. 3.1). That small addition, however, makes a dramatic difference in the stability of RNA. DNA can be left at room temperature for short periods with little fear of degradation. RNA degrades so easily that it is normally stored at $-80°C$. The difference in stability of these two information molecules relates directly to their function in the cell. DNA is the long-term repository for cellular information, whereas one of the key roles for RNA is to serve as a short-lived intermediary, carrying the information from DNA to the machinery that synthesizes proteins. The DNA in the genome can be thought of as the hard drive of a computer, which contains in its memory all of the programs necessary to run the computer. For each cell of an organism, a subset of the information contained in the genome is used. This is equivalent to what happens in a computer when one or more programs are loaded into short-term memory (or RAM).

3.2 Why Control the Amount of RNA in a Cell?

Cells need to regulate both the types and quantity of proteins they make. A hair cell makes proteins that form hair, whereas a heart cell makes proteins that form

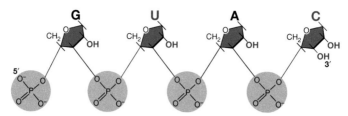

Figure 3.1. The structure of RNA. RNA molecules appear similar to single strands of DNA except that they have an oxygen and hydrogen (OH) where DNA has only a hydrogen, and they use the base uracil (U) instead of thymine (T) (compare with Fig. 2.2).

cardiac muscle. If the hair cell were to start to make cardiac muscle or vice versa, the organism would be in deep trouble. However, to make a hair or heart muscle requires several different proteins, and their proportions have to be right or there is again trouble. In a similar fashion, bacteria respond to new food sources by making new proteins to digest the food. To regulate what type, where, and how much protein is made, cells rely heavily on controlling the synthesis of **messenger RNAs** or **mRNAs** (Fig. 3.2), which relay the code for a protein from the DNA where it is stored to the factories in the cytoplasm that make proteins (see Chapter 5). The cells do this primarily by regulating the initiation of production of mRNA made from the information in individual genes.

In addition to mRNAs, there are several other types of RNA molecules that are made from DNA-encoded instructions but do not code for proteins. Among these are structural RNAs, such as ribosomal RNAs, whose role is to help provide structure to the protein synthesis factories called **ribosomes**. Another type of noncoding RNA, called **transfer RNA (tRNA)**, is involved in reading the genetic code in the mRNA during protein synthesis. Relatively recently, a host of new small RNA molecules has been discovered. These include **micro-RNAs (miRNAs)** and **short interfering RNAs (siRNAs)**, which can directly or indirectly regulate the production or abundance of specific mRNAs or protein production from mRNAs.

3.3 How Is mRNA Made?

The key enzyme for production of RNA is **RNA polymerase**. Using DNA as a template, its job is to build a chain of RNA that is precisely complementary to the DNA. It requires four bases of its own, three of which, A, C, and G are very similar to DNA bases (except for the additional oxygen and hydrogen). Instead of T, however, RNA polymerase extends the growing chain by using

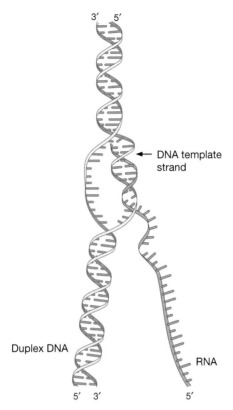

3′   5′

← DNA template
strand

Duplex DNA

RNA

5′   3′   5′

Figure 3.2. RNA is synthesized by RNA polymerase in order to form a copy of the template strand of the DNA duplex.

a base called **uracil** (U) whenever it encounters an A in the DNA strand. The act of adding bases to the growing RNA strand is called **transcription**.

When RNA polymerase encounters double-stranded DNA, it must first bind tightly to the DNA and then force the two DNA strands apart so that it can start making RNA. The process of opening up the DNA is called **melting** (Fig. 3.2). The RNA polymerase then proceeds down one of the two strands of DNA. The strand it "reads" is called the **template strand**, and thus the RNA sequence it produces is complementary to the sequence of the template strand (Fig. 3.2). The process of adding additional bases complementary to the DNA is called **elongation**. As with the synthesis of DNA, RNA synthesis is normally in the 5′ to 3′ direction (Fig. 3.2).

Once the mRNA is synthesized, it is usually modified by the addition of a string of adenosine monophosphates at the 3′ end. These are known as **poly(A) tails** and are a hallmark of mRNAs in **eukaryotic** cells (i.e., those containing a nucleus, as found in all higher organisms such as plants and animals, as opposed to **prokaryotic** cells such as bacteria, which do not contain a nucleus). Two other modifications of mRNA are unique to eukaryotic cells. First, at their 5' end, they are chemically modified to reduce nonspecific degradation. This process is called **capping**. Second, as described in Chapter 2, eukaryotic mRNAs usually contain more sequence information than is needed to code for a protein—the **primary mRNA** transcript contains both introns and exons. While the primary mRNA transcript is being made, a protein complex splices out the introns, joining the exons together to form the **spliced mRNA**. The abundance of spliced mRNA that is produced in a cell is referred to as the level of **gene expression**, which is a function of the rate of synthesis of new mRNA and of its degradation.

RNA molecules frequently have regions of their sequence that are complementary to other parts of the same molecule. In solution, these complementary regions can base-pair to form **RNA secondary structure**. As we will see, secondary structure plays a crucial role in the function of the tRNA molecules that carry amino acids to the messenger RNA during the formation of proteins—a process known as **translation**.

3.4 How Is the Amount of mRNA in a Cell Controlled?

What determines whether RNA polymerase will initiate synthesis of a particular RNA is the presence or absence of proteins called **transcription factors** that bind to the DNA, usually upstream or 5' of the gene that will be transcribed. Most of these proteins bind to specific sequences in the DNA, known as *cis*-elements. The term *cis*-element is derived from the fact that the sequences are on the same piece of DNA as the coding sequences, as opposed to being in *"trans,"* which would indicate that they are elsewhere (e.g., on another chromosome).

Two French scientists, Jacques Monod and François Jacob, won a Nobel Prize in 1965 for their insights into the regulation of mRNA synthesis in bacteria. When they added milk sugar—lactose—to the media in which the bacteria grew, they noticed that the bacteria began to make a protein—β-galactosidase—which breaks down lactose. They discovered that the mechanism by which the production was regulated was remarkably simple. A protein sits on the DNA just upstream of the starting point for the production of the mRNA that codes for β-galactosidase (Fig. 3.3). As long as lactose is not present, this protein prevents production of the mRNA. However, when lactose enters the bacteria, it causes the protein to change its shape, which in turn causes the protein to fall off from the DNA. This allows RNA polymerase to bind to

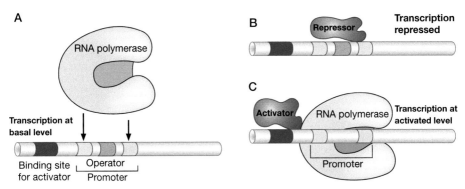

Figure 3.3. (A) In bacteria, RNA polymerase recognizes and binds to a region upstream of the start of transcription called the promoter. (B) When a repressor molecule binds to the operator site, it prevents RNA polymerase from binding to the promoter and initiating transcription. (C) Binding of an activator molecule that interacts with RNA polymerase can cause an elevated level of transcription.

the DNA and start synthesizing the mRNA coding for β-galactosidase (Fig. 3.3). The protein that prevents RNA production is called a **repressor**.

Monod and Jacob believed that this process of repressing gene expression until it was needed was so simple and elegant that it must be the general rule. In fact, both in bacteria and in other organisms, the converse process is more common in transcription control. RNA production generally does not occur until a protein binds to the DNA. This is called activation, and the protein that binds to the DNA is called an **activator** (Fig. 3.3).

The *cis*-elements in DNA that bind to repressors, activators, and RNA polymerase are usually located upstream of the sequence that encodes the mRNA. This upstream region is called the **promoter**. Promoters differ dramatically in their organization from bacteria to yeast to mammals. In bacteria, all of the *cis*-elements that regulate transcription are normally close to the transcription start site (Fig. 3.3). In yeast and plants such as *Arabidopsis*, the *cis*-elements are more spread out but usually still within a few kilobases of the start of transcription. By contrast, in mammals and other animals, the *cis*-elements can be as far away as several hundred kilobases.

How can proteins bound to DNA several hundred kilobases away from the transcription start site influence transcription? This is still not entirely understood, but it is generally believed that DNA is flexible enough to bend, which brings together proteins that are bound far apart (Fig. 3.4). In eukaryotes, the regions of DNA to which one or more transcription factors bind, thereby regulating transcription, are called **enhancers** (Fig. 3.4). The bound transcription factors are thought to help RNA polymerase start transcribing RNA (Fig. 3.4).

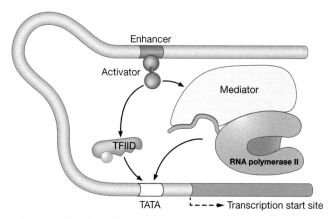

Figure 3.4. In eukaryotes, binding of RNA polymerase to DNA is facilitated by binding of the transcription factor TFIID to the TATA box that lies a short distance upstream of the transcription start site. Activator molecules frequently bind to DNA at some distance from the TATA box and are thought to interact with RNA polymerase through bending of the DNA and through an intermediary protein complex called mediator.

In fact, it is not precisely known how the binding of a set of transcription factors signals RNA polymerase to start transcribing mRNA. What is known is that, in eukaryotes, a massive complex of proteins is needed to initiate transcription. These **general transcription factors** were discovered through efforts to get transcription to occur *in vitro*. By adding different proteins extracted from cells to a mixture of DNA and nucleotides, scientists worked out which ones are able to facilitate synthesis of RNA. Proteins able to produce a low level of RNA, or **basal transcription**, are the general transcription factors. One of these general transcription factors, **TFIID**, binds to a sequence found ~35 bp upstream of where the transcript will start (Fig. 3.4). This region often has the bases T-A-T-A and thus goes by the name of "**TATA box**". Once TFIID binds, the other proteins attach themselves in a tightly defined temporal and spatial order, allowing RNA polymerase to be poised to begin synthesis of a new RNA strand. Another protein complex known as **mediator** acts as an intermediary between the transcription factors bound to enhancers and the general transcription factors (Fig. 3.4).

3.5 How Do Transcription Factors Activate Gene Expression?

Perhaps the best-understood aspects of transcriptional regulation are the transcription factors themselves. Most transcription factors are modular, with a domain that binds DNA and another domain that is involved in transcriptional

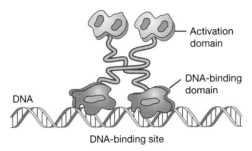

Figure 3.5. Transcription factors frequently bind to DNA as dimers and are made up of modular units, with one domain of the protein binding to the DNA and another domain interacting with RNA polymerase and serving to activate transcription.

activation (Fig. 3.5). The precise nature of the **DNA-binding domain** is what generally distinguishes one family of transcription factor from another. This region is usually highly conserved across evolutionary time, probably because of the constraints imposed upon the interaction of protein and DNA. One of the best-studied types of transcription factor, the homeodomain factors, interacts with DNA through binding to sequences of bases in what is called the **major groove** (Fig. 2.3) of the DNA helix.

In bacteria, a change in the nutrients available in the environment causes a rapid change in the mRNAs that the bacteria produce. Frequently, this is controlled by one or a small number of transcription factors. By contrast, in animals and plants, most transcriptional regulation is controlled by a large number of transcription factors. Estimates place the average between 5 and 15, and these can be bound at the 5' or 3' end of the gene, as well as within introns. Further complexity arises from the fact that many transcription factors bind as dimers (Fig. 3.5), and these can be heterodimers in which two different members of the same transcription factor family bind together. Each heterodimer can bind a slightly different DNA sequence and can interact somewhat differently with the RNA polymerase complex. Exactly how the binding of different factors, some activators, some repressors, some monomers, some homodimers, others heterodimers, controls the expression of specific mRNAs remains an open question. It is generally viewed as a probabilistic process, in which the binding of particular combinations of transcription factors either increases or decreases the likelihood of transcription initiation.

In the early days of molecular biology, it was widely believed that expression of a specific protein, such as one needed for cardiac muscle, would be controlled by activators expressed exclusively in heart cells. Although there are a few cases where cell-specific expression of a protein is indeed controlled by a single transcription factor expressed only in that cell, the far more prevalent situation is that cell-specific expression is regulated by combinations of transcription factors, no one of which is expressed exclusively in a particular cell. The same is true for responses to specific environmental stimuli, where a combination of transcription factors is usually required to bind to the DNA at the same time. This situation is analogous to the genetic code and has been called the *cis*-**regulatory code**.

3.6 What Is the Role of Noncoding RNAs in Regulating Gene Expression?

The first evidence for a role for small RNA molecules in regulating eukaryotic gene expression came from a genetic analysis of the roundworm *Caenorhabditis elegans*. It was initially thought that these small RNAs were an oddity found only in lowly worms. The availability of the human and other genome sequences allowed researchers to look for the signatures of miRNAs, which are 21–24 bases in length and are made from a longer precursor molecule. It rapidly became apparent that the human genome harbors thousands of potential miRNAs. Since their original discovery, these noncoding RNAs have been implicated in a large number of physiological processes and disease states. In plants, miRNAS have been shown to play key roles in developmental processes and in responses to abiotic stress.

Precisely how miRNAs are generated and how they function is still being worked out. Specific enzymes recognize the secondary structure of the precursor RNA and cleave the miRNA (usually 21 bases) from it. The single-stranded miRNA then targets an mRNA molecule that is more or less complementary to the miRNA sequence. The degree of complementarity in the binding determines what happens next. In animals, the binding is generally not highly complementary and a block to translation occurs. In plants, binding is usually quite complementary, leading to cleavage of the mRNA. In fact, in plants and animals, both types of miRNA interactions can and do occur. Thus, small RNAs primarily regulate gene expression at post-transcriptional stages.

3.7 What Is the Role of Global DNA Structure in RNA Expression?

Although the binding of transcription factors to DNA is important for the initiation of transcription, it is far from the whole story. Recent work has revealed a crucial role for the global structure of DNA in regulating gene expression. DNA is rarely in a naked form. In eukaryotic cells, DNA is wound around groups of proteins called **histones**. The histone complexes are called **nucleosomes** and are located at periodic intervals along the DNA (Fig. 3.6). Higher-order structures are formed

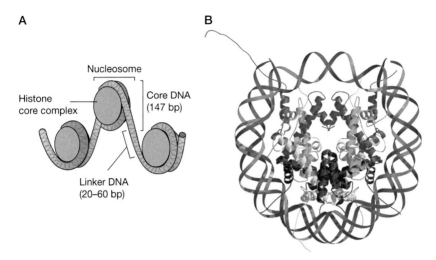

Figure 3.6. Nucleosomes comprise several different types of histone molecules (*A*), which DNA wraps around, separated by short intervening linker stretches. (*B*) Structural studies show that the histones form a core complex of eight proteins that interact closely with the spool of DNA.

as the DNA associated with nucleosomes (collectively known as **chromatin**) folds back on itself. The presence of chemical modifications on particular amino acid residues of specific histones in the nucleosome correlates with gene expression. For example, addition of an acetyl group by an enzyme called histone acetyl transferase correlates with increased gene expression. Another example is the addition of a methyl group to the ninth lysine residue in one histone, which is almost always found when a gene is expressed. By contrast, methylation of the fourth lysine correlates with silencing of a gene. Proteins that perform these modifications are called **chromatin-modifying enzymes** and form a **chromatin-remodeling complex** (Fig. 3.7). What is still unclear is how the modification of histones in nucleosomes is orchestrated with binding of transcription factors to specify when and where a particular RNA will be expressed. Because these modifications act without changing the genetic code, they are referred to as **epigenetic.**

   Another type of epigenetic regulation occurs when the DNA itself is modified by the addition of methyl groups to particular bases such as C or A. As with histone methylation, there is increasing evidence of correlation between specific patterns of DNA methylation and gene expression. In general, transposable elements are highly methylated, which is thought to keep them transcriptionally quiescent so that they will not hop around in the genome. There is also evidence that small siRNAs play a role in directing DNA methylation to specific

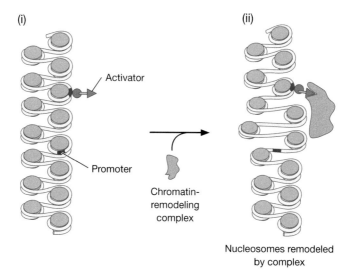

(i)   (ii)

Activator

Promoter

Chromatin-
remodeling
complex

Nucleosomes remodeled
by complex

Figure 3.7. The role of chromatin remodeling complexes is to interact with activators and displace nucleosomes, thus facilitating transcription. The activator binds to both the DNA and a chromatin-remodeling complex at a distance to the promoter (i), causing the solenoidal nucleosomal structure to be remodeled in the vicinity of the promoter (ii). This facilitates subsequent binding of the transcriptional apparatus to the promoter.

regions of the genome. Many questions remain as to precisely how epigenetic modifications are controlled and to what extent they are important in orchestrating gene expression.

3.8 How Do We Measure RNA Concentrations?

Because RNA levels are tightly controlled, measuring the concentrations of RNA found at a particular time and in a specific cell or organ can provide important information as to the function of the genes that encode them. In addition, the profile of RNAs found in a particular cell can be used as a means of disease diagnosis. For example, different types of cancer have been shown to have different RNA profiles.

For many years, RNA concentrations were measured through hybridization (pairing) of RNA to DNA. Hybridization is based on the affinity that the nucleotide adenine has for thymidine (or uracil) and that cytosine has for guanine. Under the right conditions, base pairing can be used to discriminate between sequences that are very similar to each other. The use of hybridization to measure RNA abundance was first developed for the technique known as **northern blotting** (Fig. 3.8).

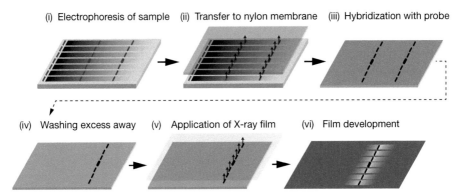

(i) Electrophoresis of sample   (ii) Transfer to nylon membrane   (iii) Hybridization with probe

(iv) Washing excess away   (v) Application of X-ray film   (vi) Film development

Figure 3.8. The steps in northern blotting include (i) gel electrophoresis of the samples of extracted RNA; (ii) transfer of the separated RNAs to a nylon membrane; (iii) hybridization with a labeled probe; (iv) washing away of excess probe; (v) imaging of bound probe with X-ray film; and (vi) development of the film and interpretation of the image.

In a northern blot, RNA samples (from different tissues, different time points, etc.) are loaded onto an agarose gel. An electric field is applied to the gel, causing the negatively charged RNA molecules to move through the gel toward the positive electrode. As mentioned in Chapter 2, electrophoresis results in separation of the population of RNA molecules according to their size (Fig. 3.8). Once sufficient separation of the RNA has been obtained, the gel is placed in contact with a membrane that will tightly bind the RNA. Capillary action is then used to transfer the RNA from the gel to the membrane, maintaining the spatial separation of the RNA.

The next steps involve the hybridization of a labeled probe to the RNA. The probe is usually DNA that has been modified so that it contains a radioactive isotope (Fig. 3.8) or an epitope that will react with an antibody. The labeled DNA is added to the membrane containing the separated RNA. Hybridization takes place under special conditions that usually include small volumes of liquid, moderately high temperatures, and low salt concentrations. These conditions are designed to allow the probe to find its complementary RNA, which remains bound to the membrane or paper, but not bind to unrelated RNA molecules. Once hybridization has occurred, the excess unbound probe is washed away and the extent of signal is deduced by, for example, use of X-ray film (Fig. 3.8).

3.9 How Do Microarrays Work?

The number of RNAs that can be analyzed in a northern blot is limited by the number of probes that are hybridized to it, which is usually one. Once genome sequences became available and we knew the sequence of every gene, the problem was how to assay the RNAs made from them all at once and not do them one at a time with northern blots.

Several different ways have been developed to determine the RNA expression levels of all genes in the genome, called **genome-wide expression analysis**. Until recently, the most commonly used technique involved the attachment of DNA to a solid support and the hybridization of labeled RNA to the bound DNA. When thousands of different DNAs are attached to a solid support, it is known as a **microarray**. Hybridization to a microarray is similar to a northern blot, with the major difference being that a microarray allows an increase in throughput by several orders of magnitude.

The first microarrays were literally printed on glass slides. Metal pins were dipped into pools of DNA and were then deposited onto the glass. The major problem was that this produced inconsistent amounts of DNA at each spot on the glass. This issue was solved by synthesizing the probes directly on a solid support, a process pioneered by the company **Affymetrix**. The synthesis is performed by **photolithography** (Fig. 3.9), a process that is used to print computer circuits—hence the name for the Affymetrix microarrays: "**GeneChips**."

On Affymetrix GeneChips, there are between 10 and 20 oligonucleotides for each gene. The choice of oligonucleotides is determined using a computer program that searches for nonoverlapping stretches of 20–25 bases that are unique in the genome.

Labeling of the target RNA (e.g., the mRNA expressed by a certain tumor cell type under investigation) is usually performed by generating a single-stranded **complementary DNA (cDNA)**, by using the enzyme **reverse transcriptase**. This enzyme is able to copy the sequence of RNA into a cDNA strand. Thus, this enzyme is able to reverse the central dogma, although it is primarily used by viruses to copy their RNA genomes into DNA. Reverse transcriptase requires an RNA template, a primer [usually oligo-dT, which binds to the terminal **poly(A)** tail of the mRNA] and sufficient nucleotides. After a first strand of cDNA has been made, the RNA is digested away using the enzyme RNase H, which is the same enzyme used to get rid of RNA primers during replication (described in Chapter 2.3). Then another enzyme is used to copy the first strand to make a second strand of cDNA. This cDNA can then be amplified by using PCR.

An alternative for labeling the target RNA population is first to make double-stranded cDNA and then to use a viral RNA polymerase to make an RNA copy, called **cRNA**. This is made possible by harnessing a special class of RNA polymerases that were first identified in viruses that infect bacteria.

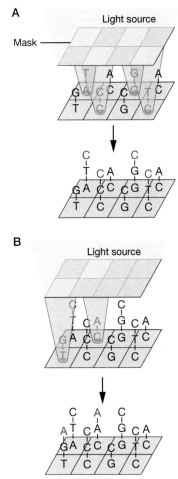

Figure 3.9. Microarrays produced through photolithography use masks to allow illumination of only those DNA strands where nucleotide addition should occur. In *A*, all sites are shielded except those where it is intended to allow addition of C nucleotides to occur. In *B*, all sites are shielded except where it is intended to allow addition of A nucleotides to occur.

These RNA polymerases require only a short recognition sequence and no additional cofactors to start making RNA. So, to make cRNA, a piece of double-stranded DNA containing the recognition site is ligated onto one end of the cDNA. Addition of the special RNA polymerase (e.g., T7 RNA polymerase) with the four bases results in production of large amounts of cRNA. Because this amplification is linear, in contrast to the exponential amplification of PCR, it can result in less distortion in the relative amounts of amplified versus starting RNA. The bases are usually modified with fluorescent labels to provide a means of quantifying the number of molecules that hybridize to the microarray.

Alpha   cdc15   cdc28   Elu

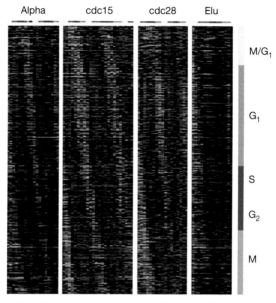

M/G$_1$

G$_1$

S

G$_2$

M

Figure 3.10. The results of a microarray analysis of cell cycle progression in yeast are here presented so that the relative level of expression of each gene is represented as a series of colors. Red indicates an increase in expression in comparison with unsynchronized cells and green indicates a decrease in expression. There were four different treatments used to synchronize the cells and these are shown as the four separate panels in the figure. At the top of each panel, the different stages of the cell cycle are indicated by colored bars. Along the vertical axis, each row is a different gene, and those with similar expression patterns are grouped together.

3.10 How Have Microarrays Been Used?

One of the first uses of microarrays was the determination of gene expression changes during the cell cycle—the phases cells go through as they prepare to divide. Of the approximately 6000 yeast genes on the microarray, more than 800 showed changes in expression at some point during the cell cycle. These genes were then grouped, or "clustered," based on when their expression rose and fell (Fig. 3.10).

3.11 What Is RNA-Seq?

Although microarrays have been used extensively to analyze genome-wide mRNA expression levels, they have been superseded by the new sequencing platforms, at least for discovery purposes. One of the main drawbacks of microarrays is that they assay only the genes that are known. As microarray production requires generating the sequence of each oligonucleotide on the array, necessarily one can only determine the expression levels of genes that have been previously identified.

Until the advent of the new sequencing platforms, it was very costly to sequence enough cDNA to compete with microarrays. A method called massively parallel signature sequencing (MPSS) was developed in the 1990s, but the cost of a single experiment was in the six figures. The dramatic drop in the cost of sequencing with the new platforms has made the cost of sequencing cDNA libraries competitive with microarrays.

The process of generating genome-wide information on RNA levels using sequencing is known as **RNA-Seq**. The steps are similar to those for microarrays. First, a cDNA library is generated. The cDNA molecules are then amplified and sequenced in the same way as genomic DNA for each sequencing platform. In addition to being able to identify RNA transcripts that have not previously been annotated, as well as noncoding RNA molecules, RNA-Seq also provides information on differential splicing (Fig. 3.11). This is primarily because each sequencing run generates hundreds of millions of reads, allowing one to find fragments that cross exon borders in different ways. Thus, several fragments might

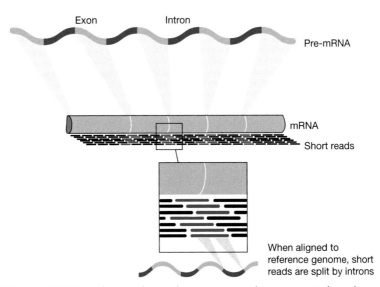

Figure 3.11. In RNA-Seq, the numbers of sequence reads are counted to determine the amount of RNA. The presence of different splice variants is indicated by sequence reads that cover both sides of a splice junction.

indicate that a particular exon was included, whereas other fragments might indicate that the exon is spliced out, suggesting that alternative splicing is occurring.

Because the number of reads generated by the new sequencers is so great, there is more **sequence depth** on each run than is needed for transcriptional profiling. This has led to the development of **multiplexing**, the addition of unique short sequences, or bar codes, to each cDNA library. This allows different libraries to be mixed and sequenced together. Thus, when the sequences are read, the bar codes identify the different starting libraries.

3.12 What Is Real-Time PCR?

Real-time PCR is another method for measuring RNA abundance. Although this approach is limited by the number of genes that can be assayed in a single experiment (e.g., hundreds of samples), it is generally considered to be highly accurate and thus serves an important role in verifying expression of specific genes detected by genome-wide methods, such as RNA-Seq.

In real-time PCR, a fluorescent signal is released at each round of amplification. In the TaqMan method, there is a fluorescent tag on one of the primers. When the polymerase arrives at the primer, it clips off the fluorescent tag and the fluorescence is recorded. The readout of a real-time PCR reaction is a set of curves that relates the level of fluorescence to the PCR cycle, with each cycle representing twice the amount of amplified product as its predecessor.

3.13 How Do We Analyze Genome-Wide Transcriptional Data?

With all existing technologies that assay RNA expression across the entire genome, there is inherent variability from experiment to experiment. Differences in the abundance of RNA are primarily due to stochastic and/or environmental factors. Variability can also arise from technical issues such as aging reagents. For these reasons, genome-wide expression experiments are usually repeated several times. There are two types of **replicates**: "technical" and "biological." As its name would indicate, a **technical replicate** uses the same input RNA sample but on different microarrays or RNA-Seq runs. The goal is to determine how much intrinsic noise there is in the experimental set up. In a **biological replicate**, by contrast, the RNA is isolated in independent experiments, usually on different days, with the aim of identifying variations that arise from some aspect of the biological process under study. By doing at least three biological replicates, statistical analyses can be performed to determine confidence levels.

It could be argued that genome-wide expression analyses were what forced molecular biologists to embrace statistics. In the early days of molecular biology, it was not unusual to hear the refrain, "If you need statistics, it means you need to do another experiment!" But, when dealing with expression analyses of more than 20,000 genes, assayed multiple times, under a variety of conditions, there is no substitute for statistical analysis.

Several algorithms based on different statistical models have been developed to identify significant differences in expression patterns from genome-wide expression analysis. There are two basic types of such algorithms: "supervised" and "unsupervised." Among the frequently used unsupervised techniques are hierarchical clustering, k-means clustering, and principal components analysis (PCA). Among the supervised clustering techniques, one that is frequently employed is the support vector machine (SVM) model.

CHAPTER 4

Cells and Cellular States

4.1 What Are Cells?

Driving through the lush vineyards that cover almost every acre of arable land in California's Napa Valley, one cannot help but marvel at how mixing grape juice with a few living yeast cells has brought unimagined wealth to this parched corner of California and inspired poems, songs, and tales that have shaped Western culture. How does yeast transform crushed grapes that would go bad in days into a nectar that improves in taste during storage for decades? To understand what yeast does when thrown into vats of grape juice, we will explore the nature of the yeast cell, what is in it, how it functions, and how it differs from the combinations of cells that make up the vine from which the grapes came and those constituting our bodies that enjoy the wine.

4.2 How Do Cells Compartmentalize Functions?

For yeast, the cell is everything. Its life is spent nourishing itself and dividing to make more yeast cells. So what constitutes a cell? All cells are enclosed by a membrane, made up of water-repellent molecules called lipids. The membrane surrounds an aqueous environment called the **cytoplasm** (Fig. 4.1). What allows lipid molecules to form membranes is that they have a long hydrophobic (water-repellent) domain and a small charged region (Fig. 4.2). In the presence of water, the hydrophobic domains aggregate and the charged "heads" of the lipid molecules interact with the water molecules. This forms

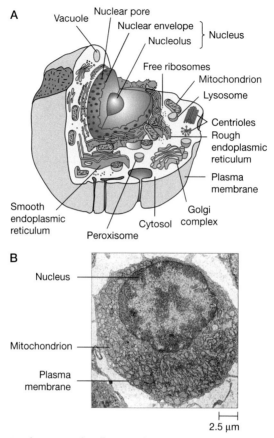

Figure 4.1. Schematic of an animal cell (*A*) and micrograph of a white blood cell (*B*), showing the nucleus and its nucleolus, the different organelles in the cytoplasm, and the membranes surrounding the cell and the nucleus.

a **lipid bilayer** (Fig. 4.2). When there are a sufficient number of lipid molecules, a spherical enclosure called a liposome will form. This ability of bipolar lipid molecules to self-organize into a membrane is thought to have played an important role in the origin of life.

Although lipids can spontaneously form a membrane, this same property poses a challenge to the cell. How can it bring in nutrients to live on? If the nutrient molecules such as sugars dissolve in water, then they will never make it past the lipid barrier of the membrane (Fig. 4.2). Cells solve this problem by having specialized protein transporters located in the membrane (Fig. 4.3). They recognize molecules such as sugars and allow them to selectively

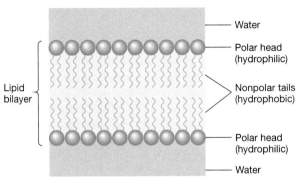

Figure 4.2. Cellular membranes are made of lipid bilayers that comprise a polar head, which is hydrophilic, and nonpolar tails, which are hydrophobic. These lipid molecules self-assemble such that the tails interact with each other and the heads interact with water molecules, thus forming a lipid bilayer.

cross the membrane. These proteins are specially designed to insert into membranes. Other proteins in the membrane serve to transfer information across the membrane. When they interact with particular molecules outside of the cell, these **receptors** alter their conformation so that a change occurs in the receptor molecule on the inside of the cell (Fig. 4.3). The change on the inside of the cell sets off a cascade of molecular interactions that allows information to flow and be amplified. Another means of getting materials into the cell is by budding off bits of membrane through a process called **endocytosis**. Viruses such as HIV hijack these processes to get into cells.

The cell is equally challenged when it wants to get molecules from inside the cell to the outside. It has the same problem of moving water-soluble molecules through the plasma membrane. To solve this export problem, cells use similar strategies to those they use to import material, including protein transporters in the membrane. But they can also package material in lipid-membrane-bound **vesicles** and send them out to the plasma membrane. When the vesicles arrive at the membrane, their lipid bilayer fuses with the lipid bilayer of the plasma membrane, which results in the contents of the vesicle being dumped outside the cell. This process is known as **secretion** or **exocytosis**.

We usually think of cholesterol as something that is harmful to our health. Indeed, too much cholesterol can cause the buildup of dangerous plaque in our arteries. But a certain level of cholesterol is essential for good health because it serves to stiffen membranes. Cholesterol is highly hydrophobic, and so it partitions into the lipid bilayer. Without cholesterol, cells would not be able to maintain their characteristic shapes.

A

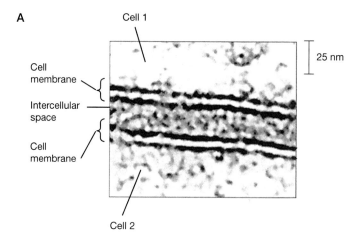

Cell 1

Cell
membrane

Intercellular
space

Cell
membrane

25 nm

Cell 2

B

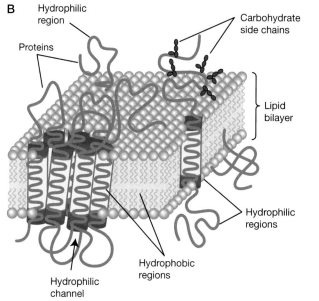

Hydrophilic
region

Carbohydrate
side chains

Proteins

Lipid
bilayer

Hydrophilic
regions

Hydrophobic
regions

Hydrophilic
channel

Figure 4.3. (A) Electron micrograph of the closely apposed membranes of two adjacent cells and (B) schematic of a plasma membrane bilayer showing how proteins are embedded in the membrane by way of domains that are hydrophobic.

4.3 What Are Organelles?

Membranes not only surround the cell, but they also form compartments within eukaryotic cells called **organelles**. As mentioned earlier, the key distinction between eukaryotic and prokaryotic cells is the presence of a nucleus. This compartment is made up of a lipid-bilayer membrane that surrounds the DNA. Because mRNA is translated in the cytoplasm, there needs to be a way to get large molecules out of the nucleus through the nuclear membrane (Fig. 4.1). This is done through complexes of proteins that form openings called **nuclear pores**. These pores also allow proteins such as transcription factors to get into the nucleus after they are translated in the cytoplasm. Once inside the nucleus, the transcription factors can bind to DNA in order to regulate gene expression. Many other types of proteins use nuclear pores to enter the nucleus, including polymerases, histones, and splicing factors. Inside the nucleus, most of the DNA is found in the **nucleoplasm**. A part of the DNA is in a region within the nucleus called the **nucleolus**, which looks darker in pictures taken with an electron microscope (Fig. 4.1). There are a large number of tandem repeats of ribosomal RNA in this region (around 200 in the human genome), and it is thought that the nucleolus acts as a small factory to churn out ribosomal RNAs for use in the translational machinery of the cell.

Within the cytoplasm, the most important organelles are arguably the **mitochondria** (Fig. 4.4), because cells depend on the energy produced in them for most of their functions. Mitochondria have two membranes—an outer and an inner—and it is within the folds of the inner membrane, known as **cristae**, that most of the energy is produced. The reactions used to produce energy from sugar by eukaryotic cells in mitochondria are described below.

Plant cells take the energy from sunlight and convert it into sugars in organelles called **chloroplasts** (Fig. 4.5). Like mitochondria, chloroplasts have two membranes, and within the inner membrane are found structures called **thylakoids** (Fig. 4.5). It is in the thylakoids that chlorophyll is found, which absorbs sunlight to produce sugar, generating oxygen as a by-product. This process of **photosynthesis** will also be discussed later.

Most eukaryotic cells have many other organelles. These include the endoplasmic reticulum (ER), in which some proteins are folded and processed; the Golgi apparatus, which sorts proteins for secretion or delivery to other organelles; lysosomes, which break down large molecules; and vacuoles in plant cells, which act as storage sites for specific molecules.

Mitochondria and chloroplasts are the only eukaryotic organelles that possess their own DNA. Long before the ability to sequence DNA, scientists hypothesized that these organelles were derived from bacteria that had invaded

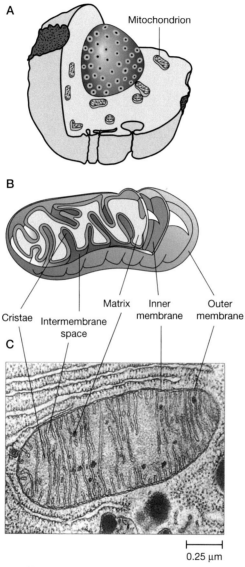

Figure 4.4. (*A*) Schematic illustrating a section through a typical eukaryotic cell showing the distribution of the cell's primary energy source—mitochondria—which are enlarged in *B*. Electron microscopy (*C*) reveals a complex internal mitochondrial structure of inner membranes that are folded into cristae, where most of the energy is produced.

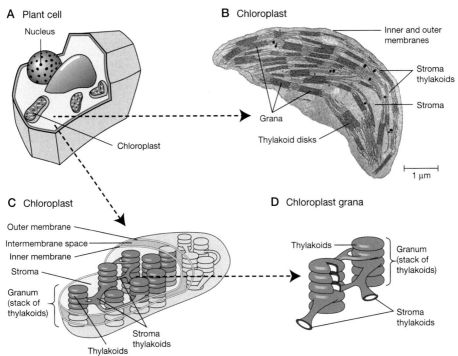

A Plant cell

Nucleus

Chloroplast

B Chloroplast

Inner and outer
membranes

Stroma
thylakoids

Stroma

Grana

Thylakoid disks

1 µm

C Chloroplast

Outer membrane
Intermembrane space
Inner membrane
Stroma

Granum
(stack of
thylakoids)

Stroma
thylakoids
Thylakoids

D Chloroplast grana

Thylakoids

Granum
(stack of
thylakoids)

Stroma
thylakoids

Figure 4.5. Schematic (A) and electron micrograph (B) of the specialized organelles called chloroplasts in plant cells. The internal thylakoid ("saclike") membranes (C) of chloroplasts are the site of the light-harvesting molecules that undertake the reactions of photosynthesis in interconnected stacks known as grana (D) lying within the chloroplast stroma.

primitive cells and formed symbiotic relationships. In support of this theory, a comparison of the genome sequences of mitochondria and *Rickettsia prowazekii*, a type of bacteria, indicated a high degree of similarity. Phylogenetic analysis suggested that the last common ancestor of *R. prowazekii* and mitochondria existed between 1.5 and 2.0 billion years ago. A similar story was used to explain the existence of chloroplasts in plant cells, the main difference being that the original invader was related to bacteria called cyanobacteria, which are capable of photosynthesis.

4.4 How Do Cells Generate Energy?

All cells need a source of energy to function and grow—the cells in the grape vines get their energy from the sun, yeast cells get their energy by fermenting the sugars in grape juice, and human cells get their energy from materials in our blood that come from the food we ingest.

The origin of all of the energy for all living things is sunlight, which is captured in reactions that take place in chloroplasts. The overall reaction combines carbon dioxide, water, and photons to give carbohydrates and oxygen. In the initial step in the process of **photosynthesis**, chlorophyll absorbs sunlight, and the energy is used to remove an electron from the first protein in a chain that begins with a protein complex called **photosystem II**. The electron is passed from an electron donor to an electron acceptor along a set of molecules that are side by side in an **electron transport chain**. In the process, protons are pumped across the chloroplast membrane, and the energy generated drives the conversion of **ADP** to **ATP** (the same molecule used in the synthesis of RNA), which is the primary energy-carrying molecule in the cell (Fig. 4.6). At the end of the electron transport chain, the electron is used to reduce a complex molecule called **NADP** to **NADPH** (Fig. 4.6). The reaction is reset by the gain of an electron for chlorophyll through splitting a water molecule, which releases oxygen (Fig. 4.6). This reaction is the primary source of all oxygen in the air.

The ATP and NADPH generated are put to work in the chloroplast to make sugars and other carbohydrates in what is called the Calvin–Benson cycle. The most abundant protein in the world is the enzyme **RuBisCo**, which captures

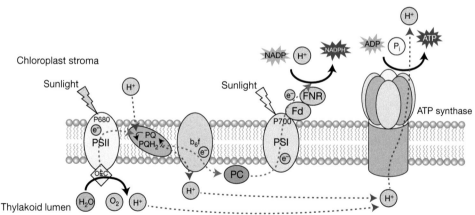

Figure 4.6. In the thylakoid membrane, a series of molecules transform the energy from sunlight into chemical energy, in the form of ATP. ATP synthesis results from the consumption by NADP of H^+ and electrons (e^-) on one side of the thylakoid membrane and by production of protons (and evolution of oxygen) from water on the opposite side. The ATP produced by ATP synthase, driven by the H^+ gradient, provides the energy needed for production of sugars. Fd, ferredoxin; FNR, ferredoxin-NADP reductase; OEC, oxygen-evolving complex; PC, plastocyanin; PS, photosystem.

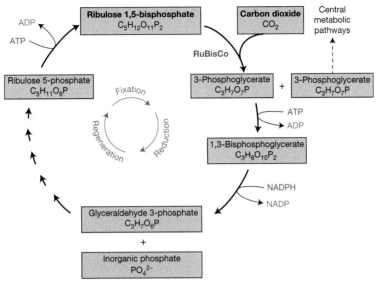

Figure 4.7. In the Calvin cycle, the enzyme RuBisCo fixes carbon dioxide by generating 3-phosphoglycerate. There are a series of steps that first involve reduction, then regeneration, of the starting substrate to complete the cycle in three phases, which are dependent on a supply of ATP and NADPH.

carbon dioxide from the air and combines it with ribulose 1,5-bisphosphate, a five-carbon sugar, to make two three-carbon sugars (Fig. 4.7). One of these molecules is then recycled to make more ribulose 1,5-bisphosphate (Fig. 4.7), whereas the other condenses with a second three-carbon sugar made in a previous cycle to form a six-carbon sugar, which ultimately will be used to make molecules such as energy-providing sucrose or the more complex carbohydrate polymer starch (Fig. 4.7).

When animals eat plant material such as fruits or vegetables, they digest the sucrose and starch, allowing their metabolism to release the energy stored in carbohydrates. Most of these chemical reactions take place in the cytosol in a process called **glycolysis**. In the first step, energy has to be put into the reaction, as the six-carbon sugar glucose is broken down to two three-carbon sugars (Fig. 4.8). This requires the energy from two ATP molecules converted to ADP. ATP molecules carry energy in the bonds that link the phosphate molecules to each other (Fig. 4.8). In the next step, the two ATP molecules are regenerated from ADP and in addition there is the reduction of two NAD^+ to NADH, which is a key cofactor required in many cellular reactions. In the final steps, the phosphate molecules are removed from two three-carbon sugars and added to two ADP molecules, generating two more ATP molecules (Fig. 4.9).

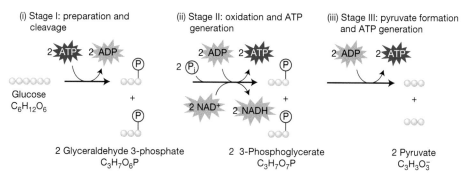

Figure 4.8. In glycolysis, a molecule of glucose is broken into two molecules of pyruvate, accompanied by the production of energy, in the form of ATP. (i) In the first, preparative stage, the glucose molecule is phosphorylated twice and split to form two molecules of the three-carbon molecule glyceraldehyde 3-phosphate, consuming two molecules of ATP. In the second stage (ii), oxidation of the two molecules of glyceraldehyde 3-phosphate to 3-phosphoglycerate occurs, with the energy conserved in the form of two ATP and two NADH molecules. In the final stage (iii), pyruvate is formed from the 3-phosphoglycerate, enabling the formation of a further two molecules of ATP.

Overall, glycolysis generates two molecules of ATP and two molecules of NADH for each sugar molecule. The cell then uses the three-carbon sugar to make even more energy. If oxygen is available, pyruvate enters the **Krebs cycle** (also known as the citric acid or tricarboxylic acid cycle), from which an additional ATP and three NADH molecules are generated (Fig. 4.10).

If oxygen is not available then **fermentation** occurs. Some organisms, such as yeast, will perform fermentation even when oxygen is available if they are placed in an environment such as grape juice where there are abundant sugars. Yeast cells generate a molecule of ethanol for each molecule of pyruvate. When muscles are overworked or when mammals have insufficient oxygen, pyruvate is converted into lactate, leading to low blood pH (lactic acidosis).

4.5 What Is a Metabolic Network?

The chemical reactions that make cellular molecules such as amino acids, sugars, and lipids, frequently interconnect and loop back on themselves. For example, the energy and cofactors produced from glycolysis and the Krebs cycle are used in numerous other reactions. The **primary metabolites** such as sugars and amino acids are used, in turn, to construct biologically active molecules such as DNA and proteins. The web of interconnecting metabolites and enzymes is called a **metabolic network**. For example, the compounds pyruvate and

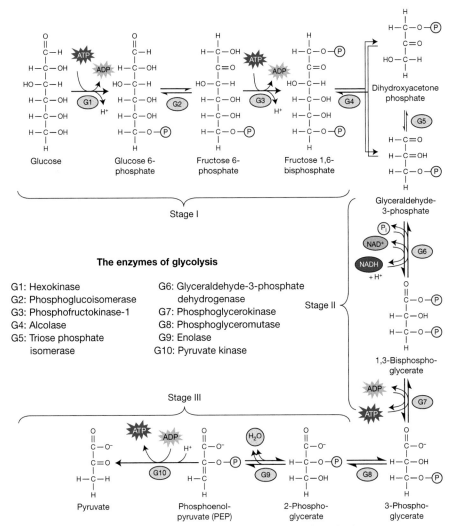

The enzymes of glycolysis

G1: Hexokinase
G2: Phosphoglucoisomerase
G3: Phosphofructokinase-1
G4: Alcolase
G5: Triose phosphate
isomerase

G6: Glyceraldehyde-3-phosphate
dehydrogenase
G7: Phosphoglycerokinase
G8: Phosphoglyceromutase
G9: Enolase
G10: Pyruvate kinase

Figure 4.9. The three-stage breakdown of glucose to pyruvate described in Fig. 4.8 is a complex process requiring 10 enzymes (here indicated as G1–G10).

2-oxobutanoate are the starting points for the biosynthesis of three amino acids—valine, leucine, and isoleucine (Fig. 4.11).

Until recently, understanding metabolic networks required painstaking biochemistry in which each step of each reaction was recreated in a test tube. Recently, the field of **metabolomics** has emerged, with the goal of identifying all of the metabolites in a cell or tissue. The challenge is to devise methods that

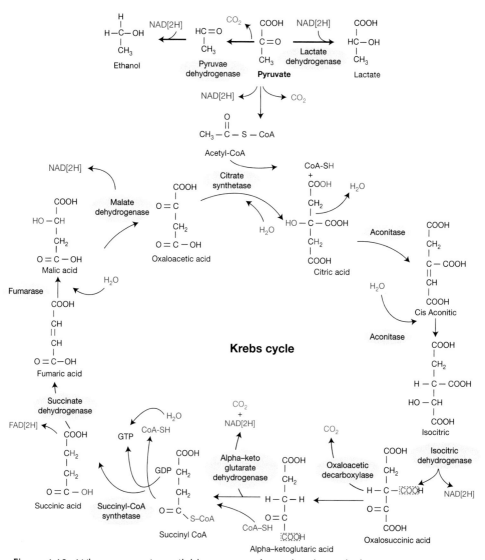

Figure 4.10. When oxygen is available, pyruvate formed at the end of stage 3 (see Figs. 4.8 and 4.9) enters the Krebs citric acid cycle (also known as the tricarboxylic acid cycle) to produce further energy and reducing molecules.

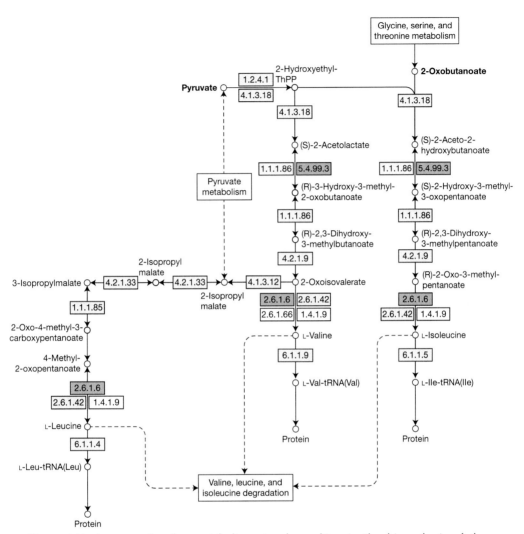

Figure 4.11. An example of a metabolic network resulting in the biosynthesis of the amino acids leucine, valine, and isoleucine from pyruvate and 2-oxobutanoate. In the boxes are the standard commission number codes, specifically identifying each enzyme-catalyzed reaction.

can identify the wide range of chemical compounds found in most cells. In one approach, the first step is to use **high-pressure liquid chromatography (HPLC)** to separate the compounds into classes (amino acids, sugars, lipids, etc.). The second step is to use **mass spectrometry** to identify the components of each class. From the mass spectrometer, a spectrum can be obtained showing peaks corresponding to most of the metabolites in the cell. The problem is that, currently, there is no straightforward means of identifying each of the compounds. The normal approach is to match a "library" of known compounds to the spectrum. At the present time, this allows the identification of fewer than 10% of the compounds present in most cells or tissues. I will discuss mass spectrometry further in Chapter 5.

4.6 How Do Cells Maintain Their Shape?

Cells use energy to grow, divide, and specialize. Cellular growth involves an increase in the size of the cell, which necessitates increases in the membrane as well as in some of the constituents of the cell. During growth, cells modify their shape by using a set of proteins that form long polymers. Together, these polymers are called the **cytoskeleton** as they form the skeletal structure of the cell (Fig. 4.12).

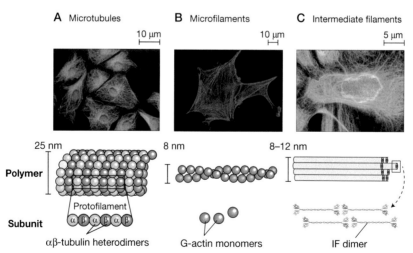

Figure 4.12. Animal cells maintain their shape through molecules that comprise the cytoskeleton, as revealed by fluorescence microscopy (*upper row*). Microtubules (*A*), microfilaments (*B*), and intermediate filaments (IFs) (*C*) are all polymers (*middle row*) that are made up of smaller subunits (*lower row*) of tubulin, G-actin, and IF proteins, respectively.

The polymers in the cytoskeleton of animal cells come in three varieties: **microtubules**, **actin microfilaments**, and **intermediate filaments** (Fig. 4.12). Intermediate filaments are primarily involved in maintaining cell shape and preserving the integrity of tissues. Microtubules are made of tubulin protein subunits and are involved in many cellular processes. In addition to maintaining cell shape, microtubules provide the motive force to separate replicated sets of chromosomes during cell division. Actin microfilaments, along with microtubules, are used for ferrying materials around the cell as well as for engineering changes in cell shape so that cells can move.

4.7 How Do Cells Divide?

Cells normally have a limit to their size. When they reach that limit, one option is to divide to make two cells. The problems facing a cell prior to and during division are daunting. Each of its constituents must be copied, and, in the case of its DNA, the doubling must be absolutely precise to maintain the integrity of the genome. The cell must then partition each half of the constituents into one each of the daughter cells. The rate of division and the number of times a cell divides are both highly regulated. If this process goes awry, the cells can die, or they can begin a runaway division process that leads to cancer. An enormous amount of research has focused on understanding how cell division is regulated. I will describe this in depth in Chapter 6.

4.8 How Do Cells Become Specialized?

The cessation of cell division is usually accompanied by the acquisition of specialized functions, a process known as **differentiation**. In a multicellular organism such as humans, cellular differentiation produces skin, muscle, blood, etc. Cells that have the potential to form a variety of differentiated cell types are called **stem cells**. These can be programmed to form any one of a number of different cell types. This programming involves changes in gene expression specific for each of the cell types. Some cells found in embryos, called **embryonic stem cells** (**ES cells**), have the capacity to differentiate into virtually any cell type. In adults, there are populations of stem cells in the gut, skin, and elsewhere that have more limited potential to form only a small number of differentiated cell types. For example, skin stem cells only form the different layers of the skin. Why embryonic stem cells are able to form an almost unlimited number of cell types, whereas adult stem cells form only a limited number, is a matter of intense research in laboratories around the world. The hope is that, if we understand the underlying regulation of cellular differentiation, we can then harness it for regenerative medicine, allowing the replacement of damaged limbs or organs. How cellular differentiation is regulated is the subject of Chapter 7.

4.9 The Concept of Cellular State

As a cell progresses from a stem cell to a differentiated cell, it passes through a series of intermediate stages. These have been given various names, depending on the tissue. For example, blood cells progress from hematopoietic stem cells to multi-potential stem cells to lymphoid progenitor cells to T cells or B cells. We can generalize this progression with the concept that cells pass through various states. A **cellular state** can be defined as the sum of all of the molecular interactions taking place in the cell at a particular stage. With currently available technology, it is impossible to know all of the molecular interactions in any cell. Recent technological advances, however, have greatly increased our ability to determine certain types of interactions. One of the key technological developments in this regard was the completion of the genome sequence of a wide range of organisms, including the human genome. From the complete genome sequence, we are able to determine the list of all of the proteins that can theoretically be produced in any cell. A second technology that has played an important role in allowing us to catalog molecular interactions is that of microarrays and, more recently, RNA-Seq, which can simultaneously record the levels of all RNAs in a cell (see Chapter 3.11). What is still lacking is a technology that identifies and quantifies all proteins and (as mentioned in Chapter 4.5) metabolites in a cell. Furthermore, beyond merely cataloging all components, we will need means of determining their dynamic interactions in living cells. For this, high-throughput imaging will play a crucial role.

4.10 High-Throughput Microscopy for Imaging Cellular Responses

The discovery of cells in plant tissues by Robert Hooke in 1665 relied on the invention of the microscope. From that time until recently, to see cells inside tissues required cutting the tissue into very fine slices so that light could pass through them. However, the invention of **confocal microscopy** in 1957 transformed our ability to image tissues.

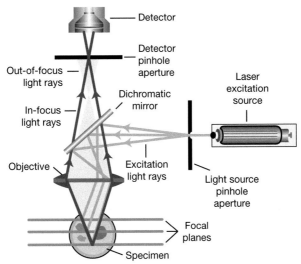

Figure 4.13. The principle of confocal microscopy is that laser light passes through the microscope objective and illuminates the specimen, where it excites fluorescent molecules. The emitted light passes through the same objective, but only in-focus light (here depicted dark blue) passes through the pinhole and is detected.

The principle of confocal microscopy is that laser light is focused through an objective onto the specimen (Fig. 4.13). The reflected light then passes back through the same objective and through a dichromatic mirror. A pinhole is positioned so that only the reflected light from the desired focal plane is transmitted to the detector (Fig. 4.13). The out-of-focus light from other focal planes is absorbed by the surface around the pinhole. Focusing on different planes allows one to "optically section" through living tissue.

Another recent advance has been the ability to automate the process of taking images with microscopes. This required bringing together computer scientists and microscope engineers. It usually is not enough to take just a series of images. More often, a computer program is also involved in analyzing the images, which can generate a feedback loop where the information in one set of images is used to home in on some aspect of the specimen for high-resolution imaging. One example is the use of automated image analysis in efforts to identify new drug candidates. The candidate compounds are placed in wells with cells. Changes in the arrangement or level of certain components are detected with specific antibodies and recorded through automated image analysis (Fig. 4.14).

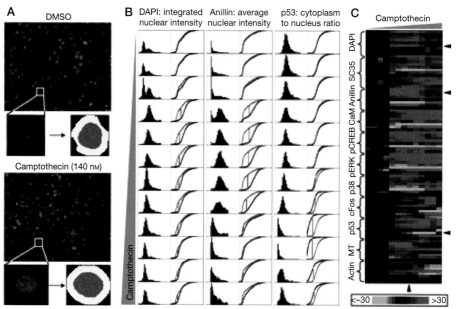

Figure 4.14. Automated image analysis (*B,C*) was used to interpret the effect of pharmaceutical agents, such as the topoisomerase inhibitor camptothecin, on different types of treated and mock-treated (DMSO) cells (*A*).

4.11 What Is the Minimum Number of Genes for a Cell to Function?

As the first genomes were sequenced, it became evident that there is a wide disparity in the number of genes found in different organisms. An intriguing issue concerned what was the least number of genes required by a self-replicating organism. As one researcher put it, this is equivalent to asking the question, "What is life?" in genomic terms.

Efforts to arrive at an answer have come from several different directions. One of the first involved a comparison between a bacterium called *Mycoplasma genitalium*, which has only 468 genes, with another bacterium, *Haemophilus influenzae*, which has 1703 genes. These two species of bacteria are only distantly related, so the reasoning was that the genes that they have in common would be the essential ones. The results of the comparison are that there should be around 250 genes in the minimal genome.

An experimental approach also started with the small genome of *M. genitalium* and tried to eliminate genes, one by one, to see which are essential. Transposons were induced to hop around the genome, and then individual colonies were analyzed to identify cases where the transposon had hopped into the middle of a gene. These were then tested to see whether the bacteria were still viable. This study identified between 250 and 350 genes that are required for viability. Remarkably, at least two other approaches have come up with similar numbers, placing the minimal genome number at between 150 and 270 genes, suggesting that a reasonable estimate is around 250. As one might expect, these include genes for DNA replication, RNA synthesis, and protein translation, as well as genes involved in energy production, cell shape determination, and cell division.

These approaches probably erred on the low side as they identified genes that were essential only under one set of environmental conditions. Another approach to answering the minimal-genome question would be to build artificial genomes and determine the exact number and type of genes that are necessary for a free-living organism to exist. Such a project is currently under way.

4.12 Is It Possible to Make a Synthetic Cell?

In 2010, Craig Venter announced that his institute had created a **synthetic cell**. There was the predictable media furor over whether it was appropriate to "create life." In fact, the scientists did not synthesize all the essential genes and place them within an artificial membrane. Instead, they started with the sequence of the genome of one bacterium, synthesized 1000-bp pieces, then ligated these together to reform the genome. They then introduced this synthetic genome into another bacterium from which they had removed the DNA. The final step was to show that the chimeric bacterium was able to grow and divide.

A long-term goal of the work on synthetic cells is to engineer them to make useful compounds such as biofuels. The idea is that metabolic pathways can be altered by introducing a set of designer genes, leading to the production of high-value products such as jet fuel. There remains a lot to be learned about which metabolic pathways to modulate and how best to engineer them to enhance the production of the desired materials or chemicals. Others are taking smaller steps in this field that is now called synthetic biology, with the goal of ultimately regulating many features of a cell.

4.13 What Is Synthetic Biology?

The field of **synthetic biology** aims to design and test synthetic networks to determine how they function when introduced into cells. This approach, called **rational network design**, uses components taken out of their native context to

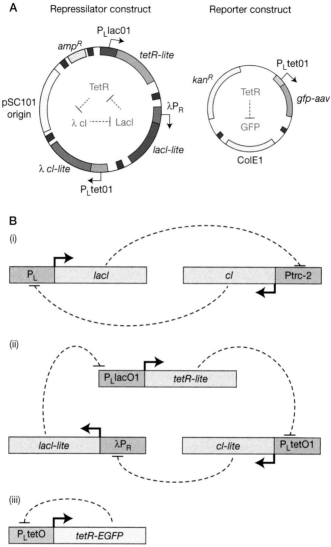

Figure 4.15. (*A*) The synthetic repressilator used a construct encoding three repressors from different organisms and a reporter. (*B*) Each repressor (i–iii) bound to and repressed the transcription of another repressor gene. By modifying the stability of the encoded proteins, the output was made to oscillate, as detected through the periodic rise and fall of the reporter—enhanced green fluorescent protein (EGFP)—whose expression was controlled by one of the repressors, the tet repressor.

create synthetic networks in cells. As with other types of engineering, one learns about the design features of the synthetic networks by perturbing them and recording how they respond. In addition to providing information on how real cellular networks function, there is the possibility of using synthetic networks to engineer new cellular behaviors that could be employed in fighting disease or in industrial applications such as manufacturing biofuels.

One of the first efforts to produce a synthetic network was called the **repressilator**, where the goal was to generate an oscillating output similar to a cellular clock (see also Chapter 6.1). This was attempted with three transcription factors from different organisms, each of which acted as a repressor. Binding sites for each transcription factor were introduced upstream of one of the other transcription factor genes so that it created a mutual-repression network (Fig. 4.15). Initially, the transcription factors were quite stable so that, once bound to their *cis*-elements, they remained bound and the whole network just shut down (Fig. 4.15Bi). The key to getting this network to oscillate was to modify the transcription factors so that they were not very stable (Fig. 4.15Bii). In this way, when one transcription factor decayed, it would allow the next transcription factor to be made until enough of the first transcription factor built-up in the cell to shut it off again. The repressilator was shown to oscillate with a period longer than a normal cell doubling time, which provided evidence that the synthetic network was able to function as designed.

CHAPTER 5

Proteins and Signaling: How Cells Regulate Information Flow

5.1 Why We Bleed Red

When you cut yourself, the red color of blood comes from a large complex called hemoglobin, which carries oxygen from the lungs to other tissues. Hemoglobin comprises four protein molecules—two each of α- and β-globin—that fold and join together to create a "pocket" for four molecules of heme and one atom of iron. Thus, for hemoglobin to function, each of the globin proteins must be synthesized as a chain of amino acids and then it must fold correctly, find its partner proteins, link to the heme molecules, and attract an iron atom. It is remarkable how often this delicate series of events works properly. Single mutations in one of the globin molecules can disrupt this process, leading to sickle cell disease, which has debilitating effects on people who carry two mutant alleles, but confers resistance to malaria to those individuals possessing one mutant allele (together with one wild-type allele).

In addition to carrying cargoes such as oxygen, proteins also serve to relay and receive signals that are essential for the cells to coordinate their activities. Within cells, proteins provide structure and serve to speed up chemical reactions. This diversity of roles might explain why, in many ways, proteins are more complex than DNA or RNA. Probably because proteins play so many different roles, they have been more recalcitrant to the introduction of high-throughput methods for their analysis. Major progress has been made, however, and more is likely in the near future.

5.2 How Are Proteins Synthesized?

Recall that the central dogma of molecular biology is that information is transferred from DNA, to RNA, to protein. Synthesis of proteins (translation) is performed by specialized complexes of RNA and protein called **ribosomes** (Fig.
5.1). These little factories are built on a scaffold of the highly abundant rRNA
molecules, which associate with a specialized set of ribosomal proteins. Ribosomes are made of two subunits, the smaller of which is the first to attach to the
nascent mRNA molecule, after which the larger one attaches itself (Fig. 5.1). In
addition to the ribosome and the mRNA, protein synthesis requires two other
players. A small species of RNA called **transfer RNA (tRNA)** (Fig. 5.2) ferries
the protein building blocks—the amino acids—to the ribosome one at a time.
The information as to which amino acid to add to the growing protein chain
is found in the codons in the mRNA. At the end of each tRNA there is a set
of three bases that are complementary to one codon. It is the remarkable specificity of hybridization that allows only the tRNA encoding the right bases
(called the **anticodon**) to match the codon in the mRNA and hence bring the

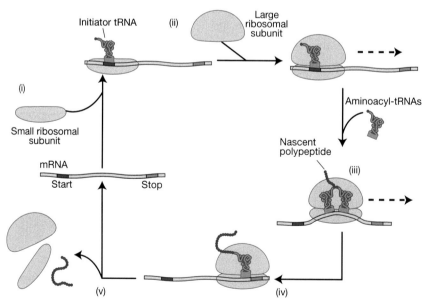

Figure 5.1. Translation is initiated when the small ribosomal subunit binds to an mRNA (i).
This is followed by binding of the large subunit to form the intact ribosome (ii). tRNA molecules carry amino acids to the ribosome (iii) to form the growing polypeptide (iv), which will
become a full-length protein (v).

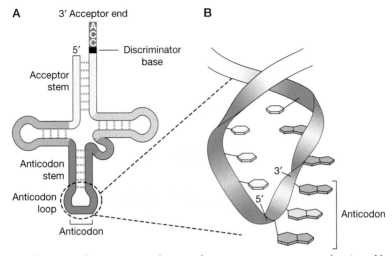

Figure 5.2. The secondary structure of tRNAs forms an acceptor arm and series of loops (A), one of which bears an anticodon (B) that is complementary to a codon found in the mRNA.

correct amino acid to the growing protein (Fig. 5.2). Thus, there is a different tRNA for each amino acid. In the ribosome, there are three positions for tRNA molecules carrying amino acids. At one end is the **amino-acyl site** where an incoming tRNA carrying its amino acid pairs with the correct codon (Fig. 5.3). Next to it is the **peptidyl site**, where the bond is made between the growing protein (also called a polypeptide) and the amino acid brought in by the tRNA (Fig. 5.3). Finally, there is the **exit site**, which, as its name indicates, is where the tRNA, having discharged its amino acid cargo, is free to leave and start the process all over again (Fig. 5.3).

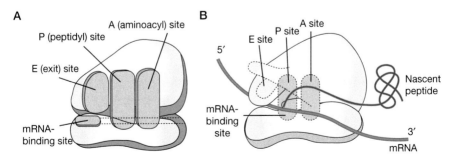

Figure 5.3. Structure (A) and function of the ribosome (B). During translation, the tRNA occupies three positions in the ribosome. When it enters the ribosome, which comprises a large and a small subunit, it is at the aminoacyl site ("A site"). The peptide bond is formed when the tRNA is at the peptidyl ("P") site, and it leaves the ribosome from the exit ("E") site.

In eukaryotes, translation is complicated by the presence of the nucleus. Instead of being able to hop on to the mRNA as soon as it starts to be transcribed, as happens in prokaryotes, the ribosomes must wait until the mRNA is exported from the nucleus into the cytoplasm. During this delay, other processing steps, which are essentially missing in prokaryotes, can take place. These include splicing of the RNA, modification of the 5′ end by capping and modification of the 3′ end by polyadenylation (see Chapter 3.3). Once in the cytoplasm, the RNA molecules are no longer tethered to the DNA, and thus the process of translation can be made more efficient. The mRNA forms a circle, allowing the ribosomes to start afresh at the 5′ end as soon as they have finished their trip around the mRNA.

5.3 How Do Proteins Form Three-Dimensional Structures?

The sequence of amino acids in any protein is called its **primary structure.** Proteins usually fold in a modular fashion, so that some regions can take on a helical form called an **α-helix**, whereas other parts might fold in a flat structure called a **β-sheet**. These modules make up the protein's **secondary structure**. The entire three-dimensional structure of a protein is called its **tertiary structure**. Proteins frequently interact with other proteins to carry out their functions, such as the interacting α- and β-globin molecules in hemoglobin. The complex containing more than one protein is referred to as the **quaternary structure** of the constituent proteins.

The stringing together of the correct sequence of amino acids is only the beginning of the long complicated process to make a functional protein. The next step is for the protein to fold upon itself to form a structure that will allow it to perform whatever it is supposed to do. If it is going to carry oxygen, for example, a protein must fold so that it forms an internal pocket (Fig. 5.4). If its job is to cleave other proteins, it must align a set of amino acids in the right spatial orientation to perform the chemical cleavage.

Unlike DNA, in which the sequence of its bases serves its primary function, protein function is dependent entirely on its three-dimensional structure. Yet, the information for the structure of a protein must be contained in the sequence of amino acids. The solution to this conundrum lies in the nature of the chemical bonds that link amino acids together. When amino acids are added to a growing protein by the ribosome, a **peptide bond** is formed between the carbon atom on the amino acid of the growing protein and a nitrogen atom on the amino acid brought in by the tRNA. This peptide bond is rigid, not allowing any rotation. However, rotation can occur around the bond formed by the

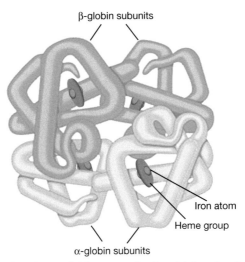

Figure 5.4. Proteins such as hemoglobin are made of multiple subunits, each of which must fold properly so that the complex can function properly.

carbon atom (the "alpha carbon," C_α) located on the other side of the nitrogen atom (Fig. 5.5). This allows the protein to adopt different three-dimensional conformations. As the protein is formed by addition of amino acids, it will begin to change its conformation in a process called **folding**. How it folds will depend on the particular sequence of amino acids and the environment in which it finds itself. For example, if the initial folding occurs in the aqueous environment of the cytoplasm, folding will occur so that the charged amino acids will interact with water molecules, and uncharged amino acids will come together in a hydrophobic pocket. Another key aspect of protein folding is that certain amino acids form bonds to other amino acids, which can be widely separated in the amino acid sequence. For example, a bond can form between the sulfur atoms in two cysteine amino acid residues, creating a **disulfide bridge**.

Although many proteins are able to find their proper conformation without aid, others get help from a class of proteins called **chaperonins**. These proteins are able to identify misfolded and certain unfolded proteins and get them to refold in the correct way.

5.4 How Do Proteins Catalyze Chemical Reactions?

Before the discovery of DNA and the attendant focus on transcriptional regulation and signaling, proteins were studied primarily for their roles in mediating the transformation of one substance into another. In most cases when a protein is involved

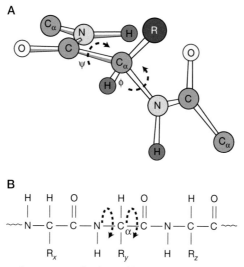

Figure 5.5. The folding of proteins is facilitated by rotation (A) that can occur between specific atoms in the chain of amino acids (B).

in transforming one substance into another, the protein itself does not change and is not consumed in the reaction. This **catalytic** activity is a hallmark of the class of proteins known as **enzymes**. Proteins that act as enzymes usually fold in such a way as to create pockets in which only specific molecules can fit (Fig. 5.6). These pockets are called the **active sites** of the enzyme. An example is the enzyme that functions in the breakdown of sucrose (common sugar) into fructose and glucose (Fig. 5.6). When a molecule of sucrose arrives in the active site, the enzyme changes its **conformation** by means of rotation along the flexible bonds of the protein. This change in conformation allows a molecule of water to enter and break a specific bond in sucrose, liberating the subcomponent fructose and glucose molecules, which are then free to circulate in the cytoplasm. This process can be repeated when another molecule of sucrose interacts with the enzyme.

When the enzyme brings the water molecule into close proximity with one bond in the sucrose molecule, it facilitates the breakdown of sucrose by lowering the **activation energy** of the chemical reaction. The creation or breaking of chemical bonds usually requires energy. By bringing molecules close together or by bending them in particular ways, enzymes can reduce the amount of energy required to get a chemical reaction to occur.

Another example of an enzymatic reaction is when a phosphate molecule (a phosphorous atom with four oxygen atoms attached) is added to a protein. Addition and removal of phosphates frequently act as "switches" to turn the

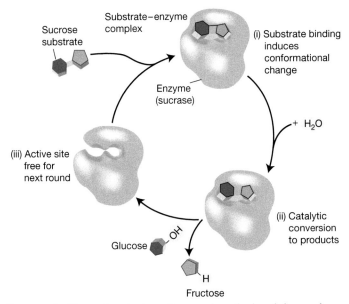

Figure 5.6. Enzymes facilitate chemical reactions such as the breakdown of sucrose into glucose and fructose through binding in the active site of the enzyme. In this example, (i) the substrate enters the active site and is maintained by weak bonds to specific amino acid residues in the active site, thus triggering a change in the conformation of the enzyme that causes the substrate to be more firmly bound; (ii) thereafter, the enzyme generates the products from the substrate, releasing them before (iii) the active site becomes free for binding another molecule of substrate again, and the cycle recommences.

activity of a protein on or off. The enzyme that catalyzes a **phosphorylation** reaction is called a **kinase**. Usually, the phosphate group is transferred from ATP to the oxygen atom on the amino acids serine, threonine, or tyrosine.

5.5 How Do Proteins Transfer Information?

Because cells are surrounded by lipid membranes, information from outside the cell can take one of two pathways into the cell. If the signaling molecule is sufficiently hydrophobic, such as the steroid hormone testosterone, it can pass directly through the membrane. However, the lipid membrane prevents the simple passage of most protein hormones such as insulin. To provide a means of allowing hydrophilic molecules to act as signals, cells embed proteins called **receptors** in their membranes. These receptors usually have three domains. There is an **extracellular domain** that interacts with the signal or **ligand**, a **transmembrane domain**, which anchors the receptor in the membrane, and a

cytoplasmic domain, which allows the signal to be transduced to the cyto-plasm, usually by means of another protein.

Communication between cells generally works as follows: (i) a cell releases a signaling molecule such as a small peptide; (ii) the peptide moves by diffusion until it encounters receptors on a cell; (iii) the peptide binds to a receptor, caus-ing a change in the conformation of the receptor protein; (iv) the alteration in protein conformation causes changes to occur in the cytoplasmic portion of the receptor, such as phosphorylation of the receptor; (v) these changes put into motion a series of events that usually amplify the original signal through a cascade of enzymatic reactions; and (vi) the final step is frequently modifica-tion of one or more transcription factors, which subsequently operate within the nucleus to regulate gene expression to respond to the incoming signal. This entire process (steps i–vi) is known as a **signal transduction pathway**. In step ii, if receptors for the signal are found on the cell emitting the signal then this creates a feedback loop. This is called **autocrine signaling**. If the signal is received by another cell, it is called **paracrine signaling**. Depending on the nature of the signaling molecule, the distance covered can be as little as between two adjacent cells or as far away as cells in different organs. Examples of long-distance signaling molecules are **hormones** and **growth factors**.

Short-distance **signal transduction pathways** have been implicated in numerous diseases. One example is the **STAT pathway** (Fig. 5.7), which has one of the simplest signal transduction processes. Binding of the ligand—a mol-ecule known as a **cytokine** such as **interferon**—causes its receptor to interact with other like receptors on the cell surface. When two of these receptors form a dimer, it activates the **Janus kinase (JAK)**, which allows it to phos-

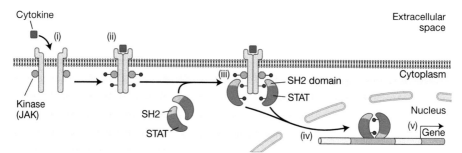

Figure 5.7. Signaling through the STAT pathway involves (i) binding of ligand (in this exam-ple, a cytokine) to its receptor; (ii) dimerization of the receptor and subsequent activation of the JAK kinase; (iii) phosphorylation of STATs; (iv) dimerization and entry of STATs into the nucleus; and (v) activation of gene expression.

phorylate tyrosine residues on the receptor. Molecules known as STATs, which have phosphotyrosine binding sites called Src-homology 2 (SH2) domains, then interact with the receptor and also become phosphorylated by JAKs. This allows the STATs to bind to other STATs to form dimers. The dimerization of STATs enables them to enter the nucleus, where they bind to specific *cis*-elements and modify transcription. Thus, this pathway encompasses only four elements: the ligand, the receptor, JAK, and the STATs (Fig. 5.7).

Contrast the relatively simple STAT pathway with the **Ras pathway** (Fig. 5.8) in which there are nine elements and at least eight steps from ligand binding to gene expression. This pathway was one of the first implicated in cancer progression. While studying several different types of cancers, investigators found that they all exhibited misregulation of signal transduction pathways that included a particular type of receptor. These receptors had the common property of phosphorylating proteins on tyrosine residues. An example of a **tyrosine kinase receptor** is the epidermal growth factor (EGF) receptor (Fig. 5.9A), which has an extracellular domain that binds to the peptide ligand EGF. There is a portion of the protein that spans the membrane and a long cytoplasmic tail that contains multiple tyrosine residues. When bound by EGF, the receptor dimerizes in a manner similar to the receptor in the JAK–STAT pathway. Dimerization of the EGF receptor causes the receptor to autophosphorylate on its tyrosine residues (Fig. 5.9B).

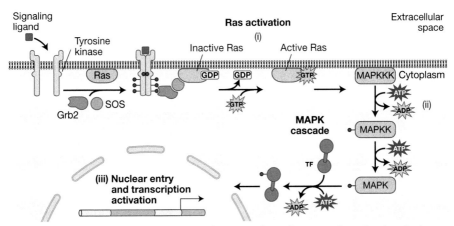

Figure 5.8. The RAS signal transduction pathway involves a large number of molecular interactions between binding of ligand to its receptor and activation of transcription. In a key step (i), the activated receptor causes GDP-bound inactive RAS to exchange GDP for GTP, become activated and henceforth initiate a mitogen-activated protein kinase (MAPK) cascade (ii) that ultimately impinges on a transcription factor (TF) that activates transcription in the nucleus (iii).

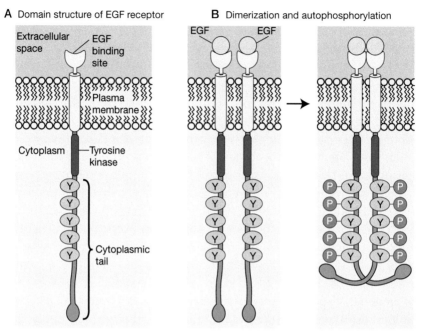

A Domain structure of EGF receptor

B Dimerization and autophosphorylation

Figure 5.9. Domain structure (*A*) and activation (*B*) of the epidermal growth factor (EGF) receptor. A common step in signal transduction in human cells involves dimerization of the ligand-bound receptor tyrosine kinase, followed by autophosphorylation (P) on tyrosine (Y) residues.

The next steps in this signal transduction pathway are set in motion by the autophosphorylation of the receptor. Three molecules are brought together at the cytoplasmic tail of the receptor: GRB2, SOS, and **RAS** (Fig. 5.8). RAS was discovered by Robert Weinberg's laboratory while searching for genes that could cause cells to behave like cancerous cells. They found a relatively small gene with a single point mutation in it that changed a glycine residue to a valine. At first, there was skepticism that they had discovered the right gene. How could a mutation from glycine to valine cause cancerous behavior? It turned out that they were absolutely correct. The seemingly innocuous change in the protein in fact caused a switch to get stuck in the "on" position. The RAS protein is a **GTPase**—an enzyme able to remove a phosphate molecule from GTP (the same GTP that is used to make RNA). When bound to GDP, RAS is in the "off" state. RAS becomes activated when it interacts with SOS, which potentiates the binding of GTP. RAS then interacts with a kinase, initiating a cascade of kinases (Fig. 5.8). These kinases are called mitogen-activated protein kinases (or **MAP kinases**). At the end of the cascade, the MAP kinase phosphorylates a transcription factor, which enters the nucleus and regulates gene expression. The mutation that Weinberg found is fairly common in different types of cancers. When GTP cannot be hydrolyzed to GDP, RAS continues to activate the kinase pathway whether or not there is ligand bound to the receptor. This constant signaling is thought to play a key role in making cancer cells unresponsive to the normal constraints that prevent them from growing in an uncontrolled manner.

Another well-characterized signal transduction mechanism also utilizes GTP-dependent switches. There are many **G-protein-coupled receptor (GPCR)** pathways, and several of them have been associated with disease states. Unlike the tyrosine kinase receptors, the G-protein-coupled receptors have multiple transmembrane domains and do not phosphorylate other proteins (Fig. 5.10). Upon ligand binding, three proteins, the α, β, and γ subunits of the **G protein complex**, associate with a cytoplasmic portion of the receptor (Fig. 5.10). The α subunit functions as a GTP-dependent switch in a similar fashion to RAS, and, when GDP is bound, the α subunit is inactive. Association with the G-protein-coupled receptor causes an exchange of GDP for GTP and activates the complex (Fig. 5.10). The activated α subunit then dissociates from the complex and initiates the subsequent steps of the signal transduction pathway through interaction with other components, usually enzymes (Fig. 5.10).

There are other signaling mechanisms that don't involve phosphorylation or GTP. Instead, these pathways use small molecules, called **second messengers**, such as cAMP (related to ATP) and inositol trisphosphate [Ins(1,4,5)P_3 or IP3], which is a breakdown product of a lipid molecule located in the plasma membrane. The IP3 signaling pathway also makes use of calcium ions (Ca^{2+}) as second messengers. Ca^{2+} is maintained at relatively low levels in most cells by actively pumping it out of the cell. IP3 binding to its receptor triggers release of Ca^{2+} from the endoplasmic reticulum. The released Ca^{2+} can bind and activate several proteins, of which the best characterized is **calmodulin**. When bound, calmodulin activates its target proteins, continuing propagation of the signal.

5.6 What Is a Signaling Network?

The distinction between a signaling pathway and a signaling network is that the latter involves multiple incoming signals and/or multiple outputs. The pathways described in the last section frequently intersect with other pathways through what is called **signaling cross talk** to form a signaling network.

An example of a signaling network involves a protein that was named for its apparent molecular mass, **p53**. Originally identified as an **oncogene**, it later

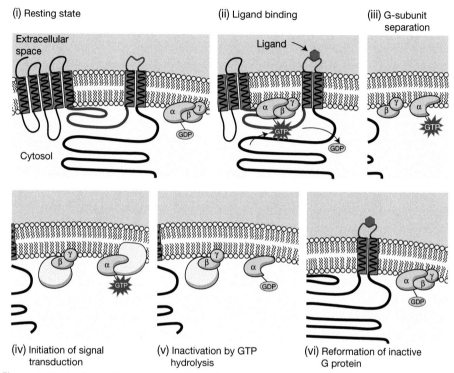

(i) Resting state

(ii) Ligand binding

(iii) G-subunit separation

Extracellular space

Ligand

Cytosol

GDP

GTP

GDP

GTP

(iv) Initiation of signal transduction

(v) Inactivation by GTP hydrolysis

(vi) Reformation of inactive G protein

Figure 5.10. In the resting state (i), G-protein-coupled receptors (GPCRs) have seven trans-membrane domains, and the G protein, comprising three subunits (α, β, γ), is bound to GDP. Upon ligand binding to the GPCR (ii), an exchange of GTP for GDP occurs on the G protein complex, which results in the dissociation of the complex (iii) and release of the α subunit. This subunit then activates or inhibits other proteins to transduce the signal (iv). For termination of the signal, the G_α subunit hydrolyzes its bound GTP to GDP (v) and the G protein subunits reassociate to form an inactive GDP-bound G protein complex (vi).

became apparent that its real role is that of a **tumor suppressor**. Oncogenes, when activated, cause cancer, whereas tumor suppressors can lead to cancer when the activity of the suppressor is reduced. In fact, about half of human cancers have mutations in the gene encoding p53. It functions as a cellular sentinel, acting on a wide range of signals to prevent inappropriate cell proliferation or to cause the cell to self-destruct.

The incoming signals that trigger p53 activity include UV light, chemicals that cause mutations in DNA, and deregulated oncogene activity. These signals activate a variety of signaling pathways that converge on kinases that phosphorylate p53 (Fig. 5.11). Once p53 is phosphorylated, it changes its conformation, which allows it to bind to DNA and act as a transcription factor. One of its targets is the

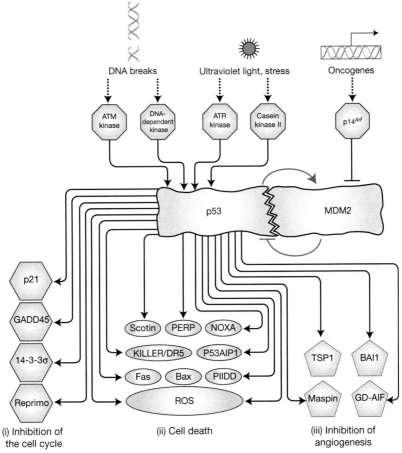

Figure 5.11. The tumor suppressor p53 integrates information about DNA integrity, UV-induced DNA damage and the activity of oncogenes to generate a decision to (i) arrest cell growth, (ii) undergo programmed cell death (apoptosis), or (iii) prevent new blood vessel formation (angiogenesis).

gene encoding **p21**. This protein acts as a "brake" on the cell cycle, preventing further cell division until repairs to the genome can be carried out. Thus, if p53 is defective, p21 will not be activated when damage occurs to the genome, the cell will attempt replication of the damaged genome and mutations can rapidly accumulate, leading to runaway cell division.

When genomic damage is beyond repair, p53 activation can lead to programmed cell death by a process called **apoptosis**. The p53 network is designed to be able to weigh the type and intensity of the incoming signals so that it can "decide" whether to just block replication (allowing time for repair) or to force the cell to undergo apoptosis. The precise mechanism by which the network makes this crucial choice is still being worked out. It is worth noting that suppression of tumors is not something that is intrinsic to the p53 protein, but is a consequence of the way the signaling network has evolved and functions.

5.7 How Are Proteins Identified in Complex Mixtures?

Because of their complexity, proteins are far more difficult to identify in complex mixtures, such as a cellular extract, than are nucleic acids. The remarkable specificity of base pairing has been the basis for most high-throughput methods for nucleic acid identification, such as PCR. No equivalent yet exists for proteins. Nevertheless, there are several approaches to high-throughput protein identification that are rapidly improving our ability to identify proteins. Foremost among these is **mass spectrometry**.

At first glance, one would not expect a mass spectrometer to be useful in protein identification, as this instrument does only one thing: It is able to measure the ratio of mass to charge of a molecule. Proteins can have many charged amino acids (such as positively charged lysine or negatively charged aspartate) in their primary structure, and thus distinguishing them by a mass-to-charge ratio would seem to be impossible. However, by implementing a series of ingenious steps, this is now a routine procedure.

The first step is to break a protein into smaller, more manageable pieces. This is usually done by treating it with a **protease**, an enzyme that cleaves proteins. The most commonly used protease is trypsin, which cuts the protein after every arginine or lysine amino acid residue. The pieces of protein then need to become ionized or have a charge applied to them. There are several different approaches, but one of the most popular is **electrospray ionization (ESI)**. ESI is commonly used with **liquid chromatography**, which produces an initial separation of the complex mixture of proteins found in most biological samples. When they exit from the chromatography column, the proteins are ionized by passing through a very fine needle and an electric field (Fig. 5.12).

An alternative to ESI, **matrix-assisted laser desorption/ionization (MALDI)**, uses a laser to vaporize the pieces of protein in a process called desorption (Fig. 5.12D). The matrix is made up of small molecules that readily absorb UV light. The protein fragments are mixed with the matrix, which turns into a crystalline

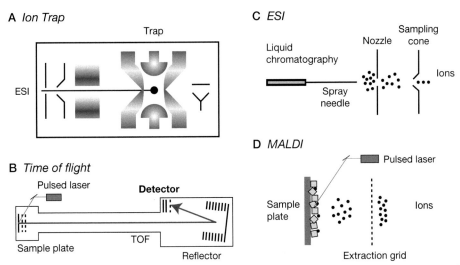

Figure 5.12. In mass spectrometry, the mass-to-charge ratio of molecules can be measured using (A) ion traps or (B) time-of-flight detectors. The molecules can be ionized by means of (C) electrospray ionization (ESI) or (D) matrix-assisted laser desorption/ionization (MALDI).

solid. When a UV laser pulse hits the matrix, the protein fragments vaporize and gain protons and thus become positively charged.

Once protein fragments are ionized, there are various means of determining the mass-to-charge ratio. For ions produced by either ESI or MALDI, the most common approach is **time of flight (TOF)**. In a TOF analyzer, a uniform electric field is applied to the ions, accelerating them down a long tube. Molecules with the same charge will have the same kinetic energy, but their velocity will depend on the mass of the molecule. Thus, by measuring the time it takes to travel a known distance, the mass-to-charge ratio can be determined. In today's mass spectrometers, there are usually tandem TOF analyzers, and the accuracy is better than 10 parts per million.

The final step is to infer the sequence of the protein based on the fragments identified by their mass-to-charge ratio. The availability of complete genome sequences has greatly facilitated this process, allowing every trypsin cleavage site to be computationally identified for every protein encoded in the genome. The mass of each of the potential protein fragments can then be calculated. Analysis of the output of a mass spectrometer requires finding the best matches between the masses measured and the masses computed from the genome. Finding several fragments from the same protein increases the confidence that a protein has been accurately identified.

5.8 How Can the Differences in Proteins between Two Samples Be Measured?

Mass spectrometry is the primary workhorse for protein identification. However, it is not possible to determine accurately the concentration of proteins in a manner similar to that of RNA-seq determining the abundance of RNAs (see Chapter 3.11). Recently, an approach called **selected ion** or **multiple reaction monitoring** was developed in which peptides are synthesized that are identical to naturally occurring peptides except that they contain heavier forms of carbon and nitrogen (^{13}C or ^{15}N) incorporated into certain amino acids. By comparing the known amounts of these labeled peptides added to a sample with the amounts measured by mass spectrometry, accurate concentrations can be determined.

As an alternative, methods have been developed to measure the relative levels of proteins, for example those in a tumor in comparison with those of normal tissue. The problem has been to incorporate a ^{13}C or ^{15}N label uniformly into the proteins of one of the samples. For single-cell organisms such as bacteria, they can be fed amino acids that contain the stable isotope. For plants or animals, the label is usually added by means of enzymatic or chemical reactions after the proteins are isolated, but this rarely results in uniform labeling. The labeled and unlabeled samples are then mixed and placed in the mass spectrometer. The difference in mass attributable to the label can be calculated so that the peaks from the two samples can be identified. Comparison of the area under the peaks for the same fragments from different samples yields the relative abundance of each.

5.9 How Are Protein–Protein Interactions Identified?

Although identification of proteins and their activities is crucial to understanding their biological roles, most proteins are found in complexes in which they interact with other proteins. Identifying the interaction partners can provide important clues as to the function of the protein.

One of the most popular approaches for identifying interacting proteins takes advantage of the modular nature of transcription factors. Most transcription factors have a domain that binds the DNA, and a separate domain that is involved in activating transcription. In the **yeast two-hybrid method**, a DNA-binding domain from the Gal4 transcription factor, for example, is fused to one protein, which is called the "bait." Another protein is fused to the Gal4 activation domain. If the two proteins interact, they reconstitute the transcription factor and can activate transcription from any gene whose promoter has a binding site for Gal4. Frequently, the Gal4 binding sites are placed upstream of two genes; one activates expression of a gene required to make an essential amino acid, allowing the yeast to grow, and the other activates a reporter gene whose activity can be visualized with a dye (Fig. 5.13).

The yeast two-hybrid method has been used both to determine whether specific proteins are likely to interact and to identify interacting partners across the whole genome. In the latter approach, all of the open reading frames (**ORFs**) in DNA that might encode proteins are fused to the DNA-binding domain in one set of yeast cells, and fused again to the activation domain in another set of yeast cells. The two sets of cells are then mated and placed under selection so that only those cells that have reconstituted a functional transcription factor can grow. From the cells that survive, the DNA of the yeast two-hybrid construct is isolated and sequenced to determine which ORFs have interacted.

Because the interactions identified by the yeast two-hybrid method occur in the context of forming a transcription factor in the yeast nucleus and not in their natural environment, there has been criticism that there are likely to be a lot of false positives and false negatives. In particular, false negatives are likely to arise when proteins require a membrane environment to interact or only interact when they are modified by some means such as phosphorylation. To address these concerns, modifications of the method have been developed, including one in which the interactions occur at the plasma membrane. Although

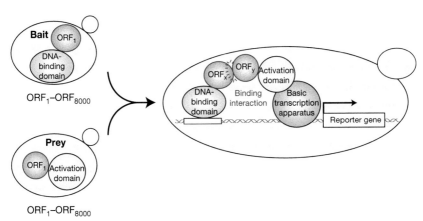

Figure 5.13. The yeast two-hybrid method is used to identify proteins that are able to interact. It is based on the modular nature of transcription factors, which comprise a DNA-binding domain and an activation domain. In the set-up stage, one set of open reading frames (ORFs) is fused to a DNA-binding domain (the "bait"), whereas another set of ORFs is fused to an activation domain (the "prey"). Only when the two protein products of the ORFs (here denoted ORF$_x$ and ORF$_y$) interact in the yeast cell will there be reconstitution of an active transcription factor that leads to expression of a reporter gene.

concerns remain, genome-wide interaction maps of many organisms, including humans, fruit flies, and *Arabidopsis,* have been generated, and the results have proved very informative for other studies.

An alternative to the yeast two-hybrid method is to use antibodies to isolate a protein, then determine what other proteins are bound to it. Antibodies are proteins made by the immune system to fight infections (among other things) and can bind in a highly specific manner to other proteins. For example, by injecting the protein of interest into a rabbit, specific antibodies can be generated. These antibodies can then be added to an extract of proteins from a source such as liver tissue. The proteins bound by the antibodies are purified away from all the other proteins by using reagents that bind specifically to the antibodies. The proteins purified in this way can then be analyzed by a method such as MALDI-TOF mass spectrometry to determine what proteins are bound to the protein "pulled down" by the antibody.

Because making antibodies to every protein in the genome would take a long time and be very expensive, an alternative is to place a **tag** on specific proteins. Tags are usually short stretches of amino acids that are recognized by specific antibodies and are unlikely to interfere with the normal functioning of the protein. In this way, the same antibody can be used for many experiments. The tag is added to the protein-coding region of the gene of interest and introduced into the plant or animal by a process called **transformation.**

5.10 Is It Possible to Identify Where Proteins Bind to DNA?

In addition to knowing which proteins bind other proteins, it would be very useful to know where transcription factors bind to DNA. This has been a goal of biologists since transcription factors were first identified, but still is only able to be inferred. There are two widely used approaches, one takes a gene-centric view and asks what transcription factors bind to the regulatory region of a specific gene, and the other takes a transcription-factor-centric view and asks which regions of DNA are bound by a specific transcription factor.

The first approach is a variation on the yeast two-hybrid technique in which the regulatory region of the gene of interest is placed upstream of a selection gene such as one coding for an essential amino acid. This is the bait. Each transcription factor (either likely binding candidates or the entire set from the genome) is then fused to a yeast activation domain. These are the prey. When bait and prey are brought together in the **yeast one-hybrid method,** the transcription factors that are able to bind to the regulatory region of the gene of interest will activate transcription and allow the cells to survive. As this method only determines that a transcription factor can bind to the regulatory region, further analysis is needed to determine whether the factor acts as an activator or as a repressor.

If the goal is to identify the genes regulated by a specific transcription factor, then the most commonly used approach is chromatin immunoprecipitation followed by sequencing (**ChIP-Seq**). The first step is to capture the transcription factor as it is bound to DNA. To do this, the sample is treated with a mixture of formaldehyde and alcohol, which causes proteins to form cross-links to DNA and to other proteins. Because the DNA is still bound by nucleosomes, it is in the form of chromatin. The next step is to break the chromatin into manageable pieces, usually 200–1000 bp in length, by sonicating the sample. Antibodies specific to a particular transcription factor or to the tag fused to a transcription factor are then used to isolate the pieces of DNA bound to the transcription factor. The cross-linking is then reversed, releasing the DNA from the proteins bound to it. The isolated DNA is sequenced and the sequences compared with the genome sequence to establish where the proteins bind.

Although ChIP-Seq has been very useful in identifying candidates for genes that a transcription factor might be regulating, it is by no means foolproof. There are usually a far greater number of potential binding regions identified by ChIP-Seq than are implicated by any other method. To date, this discrepancy is not well understood. It is possible that ChIP-Seq identifies binding sites that are used only under special conditions. Alternatively, transcription factors might bind to many sites in a fleeting or unproductive manner and to only a relatively few sites in a way that leads to activation or repression of transcription.

5.11 Building a Regulatory Network

If one can determine which transcription factors regulate a set of target genes and among those genes are additional transcription factors, then one starts to build a **gene-regulatory network (GRN)**. The nodes and edges of these networks have been identified in a number of different ways.

One of the most famous GRNs describes the factors involved in regulating genes during sea urchin embryogenesis. This was generated in a painstaking fashion by identifying candidate transcription factors then increasing or reducing their expression and determining the effects on potential target genes, as described in Chapter 7. More recently, GRNs have been constructed by combining ChIP-Seq (or its predecessor ChIP-chip) data with genome-wide expression data from RNA-seq or microarrays. The rationale is that if you know what genes are expressed at the same time and you know what transcription factors bind to these genes, then you can deduce the structure of the network. An early example of this type of effort focused on the yeast cell cycle because there were several independent data sets of microarray expression and the targets of many transcription factors that regulated cell cycle genes were known.

Within GRNs, researchers identified several **network motifs** that occurred at a frequency higher than would be expected by chance. A straightforward example is an autoregulatory loop in which a transcription factor binds and regulates its own expression. A more interesting motif is a **feedforward loop** (Fig. 5.14). This is when a transcription factor turns on another transcription factor and the two together are needed to activate or repress a downstream target. Uri Alon has proposed that one of the functions of feedforward loops is to buffer input information so that responses only occur when there is sustained input. For example, a nutrient source for bacteria might only appear in its environment for a short time. It would not be worth producing all of the enzymes necessary

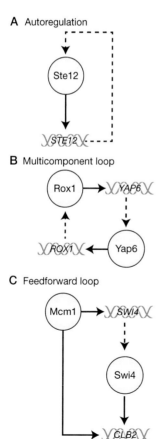

Figure 5.14. In gene-regulatory networks, some motifs are overrepresented. These include (A) autoregulation, when the transcription factor controls the expression of its own gene; (B) the multicomponent loop, in which one transcription factor regulates the expression of another transcription factor, which regulates the expression of the first factor; and (C) the feedforward loop, in which a transcription factor regulates the expression of another transcription factor and, together, both factors regulate the expression of a third gene.

to utilize the nutrient unless it remained present for a while. In a feedforward loop, the first transcription factor would sense the presence of the nutrient but require continued production of the second transcription factor to activate the enzyme genes. This would ensure that the enzymes are not made unless there is continual production of the first transcription factor.

5.12 How Can Proteins Be Purified?

To determine the structure of a protein or to analyze its activity usually requires large amounts of the protein. Before scientists gained the ability to manipulate DNA, proteins were purified from readily available sources in which they were thought to be abundant. For example, insulin was purified for use in treating diabetes from thousands of pig pancreases. With the advent of **recombinant DNA technology**, which provided the ability to manipulate and introduce DNA into bacteria, there was the possibility of reprogramming the bacteria to make large amounts of a specific protein of interest. There are many pharmaceutical agents, known as **biologicals**, that are purified from bacteria or other cell types today for use as therapeutics.

A major issue in having bacteria synthesize large amounts of a foreign protein is finding a way to purify the protein away from the bacterial proteins. The solution is the use of a tag like that described above for identifying protein interactors. One of the most popular tags for protein purification is the addition of a series of histidine amino acid residues known as a **His tag**. At high pH, the histidines are negatively charged and bind to positively charged nickel atoms immobilized on a column. By lowering the pH, the histidines become positively charged and can be removed from the column. Other tags are short stretches of amino acids for which an antibody exists. In those cases, purification is performed over a column containing the specific antibody.

5.13 How Are Protein Structures Determined?

Once proteins are purified, it is possible to determine their precise three-dimensional structure. The most commonly used method to determine protein structure is **X-ray crystallography**. As its name suggests, the technique involves shining an X-ray beam through a crystal. However, to produce crystals from the irregular structures of proteins is more art than science. Nevertheless, a large number of proteins have been successfully crystallized. The X-ray beams used in most applications are the by-products generated when electron beams are deflected in synchrotrons. These **beam lines** have extremely high energy, and their use has greatly reduced the time necessary for X-ray crystallography. The protein crystal is slowly rotated in front of the beam line, and images are captured at each angle. The images consist of dots that are formed when the X-ray beam is interrupted by the electron cloud surrounding an atom. The combination of dots is called the dif-fraction pattern, which is formed as the waves of X-rays either reinforce or cancel each other out. These diffraction patterns are analyzed using Fourier transform methods to identify the placement of every atom in the protein. From a well-formed crystal, the resolution of X-ray crystallography can be as high as a single angstrom and therefore can reveal the atomic arrangement within a protein. A shortcoming of X-ray crystallography is that it only works on well-formed crystals, which are a far cry from the natural state of proteins in a cell.

Nuclear magnetic resonance (NMR) is able to provide structural information on proteins in an aqueous environment. It uses powerful magnets to generate electromagnetic energy at radio frequencies, which is absorbed by nuclei in the protein molecule. Different nuclei emit spectra of different shapes depending on their proximity to other nuclei, which in theory can allow researchers to deter-mine not only the types of atoms in a protein molecule but also the distance between them. The primary drawback has been that these spectra become increasingly complex with protein size such that most NMR structure determina-tions have been performed on small proteins (fewer than 350 amino acids) or pro-tein fragments. However, recent advances have allowed the determination of the structure of much larger protein complexes. Nevertheless, the resolution is usu-ally lower than that obtained through X-ray crystallography.

Protein structure determination plays an important role in drug discovery. Many important drug targets are receptors that normally reside in the plasma membrane. To extract these proteins requires the use of detergents, which unfortunately then inhibits the formation of good crystals. One approach to address this issue has been to reconstitute the receptors in lipid bilayers, which can be analyzed by NMR. This is still challenging and has had limited success to date. An alternative, despite being at a substantially lower resolution, is to use cryo-electron microscopy. This remains an area of active investigation. To date, protein structure determination has been mainly a cottage industry, with each research group working on one or a small number of proteins at a time. How-ever, the field of **structural genomics** has as its goal to determine the three-dimensional structure of every protein in the genome. Efforts are being made to automate crystallization and data analysis. There are also parallel efforts to model protein structure from the primary amino acid sequence. Both efforts have had limited success, so far.

CHAPTER 6

Networks Controlling Biological Oscillators

6.1 What Are Biological Oscillators?

Stepping off a plane in a foreign country usually means adjusting to a different time zone. One's body wants to eat and sleep at the times it is used to rather than when the locals do. Fortunately, this circadian cycle can be reset after a few days, but this means that one has the same problem on returning home.

Another type of cycle occurs every time a cell divides. The cell must double all of its contents, including its DNA, and then partition them into two newly formed cells. Remarkably, both oscillating processes are governed by molecular networks that share common features—they both depend on clock-like molecular oscillators that have autoregulatory feedback loops. Other biological oscillators that are beginning to be described in terms of networks include a clock-like process for generating somites (the precursors to vertebrae) and one for positioning plant lateral roots.

6.2 What Is the Cell Cycle?

Cancer is among the most dreaded diseases. It has proven remarkably intractable, in large part, because it is a malfunctioning of a basic process—cell division. The most common treatment for cancer, chemotherapy, attempts to stop cell division, and, in so doing, stops all cells from dividing, such as those needed for hair growth. The quest to find cures for cancer focuses on gaining a better understanding of the process of cell division.

For a cell to divide, it must first make copies of all its important molecules, particularly its DNA, then partition the copies so that the two daughter cells each get enough to function. These two major steps, duplicating the cellular contents and segregating them into the daughters, are performed at different times. Not surprisingly, there is an intricate system that ensures that everything has been duplicated before allowing division to occur. The steps occurring during cell division are known as the **cell cycle** and are very similar in organisms ranging from yeast to man.

The first phase of the cell cycle, called G_1 (Fig. 6.1), is the time of preparing to make copies of all key molecules. The next phase, **S**, is when a copy of the chromosomal DNA is synthesized, along with replication of other molecules. The G_2 phase is when the cell "pauses" to make sure that the genome has been copied accurately, and finally the parental cell partitions the DNA into the daughter cells during **M phase** (for **mitosis**). The act of segregating the DNA and the formation of two new cells is called **cytokinesis**.

Failure to perform any of these steps correctly can cause arrest of the cell cycle and the inability to progress further along the cycle. In contrast, if cell division proceeds too quickly or does not stop when sufficient cells have been made to populate a tissue or organ, this can lead to the initiation of a

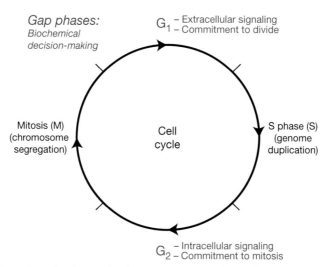

Figure 6.1. The cell cycle of typical eukaryotic somatic cells and yeast. The cycle comprises four phases: the first gap phase (G_1), which ends in a commitment to undergo the synthesis (S) phase, which is followed by the second gap phase (G_2), which concludes the three interphase stages and is followed by mitosis (M), in which the cell separates its genetic material and divides in two.

tumor. To understand how the cell cycle is controlled, researchers have used genetic screens to identify mutations that disrupt the different cell cycle phases.

6.3 What Is a Genetic Screen?

In a **genetic screen**, one looks for mutants with altered **phenotypes**. A phenotype is nothing more than a change of any sort when compared with the normal or **wild type**. A phenotype can be a different eye color or a different blood type or any attribute that one is interested in. The basic idea of genetics is that, if you can identify a mutant with an altered phenotype, then the gene that has been mutated plays a role in producing the normal phenotype.

Although there are naturally occurring mutations that can be studied, most genetic screens start by inducing mutations with chemicals such as **ethyl methanesulfonate (EMS)** or radiation, which cause changes in the DNA. The progeny of the mutated individuals are then screened for the phenotype of interest.

In the early 1970s, Lee Hartwell performed a genetic screen to identify genes that are important for cell cycle progression. He used baker's yeast (*Saccharomyces cerevisiae*) because it divides by making a bud—a protrusion from the mother cell that grows larger throughout the cell cycle until it pinches off to form the daughter cell. The size of the bud is a good indicator of the phase of the cell cycle. However, there was a small problem that needed to be overcome. If mutations caused the yeast to stop in the cell cycle, how would they ever be analyzed? Once the cell cycle stops, no more progeny are made— thus, the strain would be lost. The solution was to use **temperature-sensitive mutants**. These mutations allow cells to divide normally at the **permissive temperature** (in this case, 23°C), but they show the mutant phenotype at the **restrictive temperature** (36°C). Temperature-sensitive mutations frequently cause the protein to be unstable at higher temperatures. Hartwell allowed the cells to divide normally at the permissive temperature, then transferred them to the restrictive temperature and photographed them periodically through a microscope. Using this genetic screen, a large number of mutants that were blocked at different points in the cell cycle were identified.

6.4 How Is the Cell Cycle Controlled?

Analyses of the mutations that disrupted the cell cycle, combined with biochemical studies, have led to a detailed picture of the molecules that drive cell division. One of the key players is a protein named **cyclin**, which binds and activates a kinase called **cyclin-dependent kinase (CDK)** (Fig. 6.2). Cyclin was originally identified in extracts from sea urchins when Tim Hunt found that there was a molecule that built up, then disappeared, only to start a new cycle. Paul Nurse identified the first CDK in a genetic screen similar to the one performed by Hartwell. For their discoveries of the basic components of the cell cycle, Hartwell, Nurse, and Hunt shared a Nobel Prize in 2001.

The cyclin–CDK interaction generates an oscillator as cyclin builds up to a level sufficient to activate CDK, which is present at constant levels. The activated CDK then phosphorylates and thereby activates several target proteins, including **anaphase-promoting complex (APC)**. The job of APC is to mark for destruction proteins that are no longer needed. One of the proteins it marks is cyclin, and this causes cyclin to be degraded, thus closing the loop (Fig. 6.3A). The period of this negative-feedback oscillator is on the order of minutes in the sea urchin egg, which drives the rapid proliferation of cells (Fig. 6.3B).

In addition to activating the APC, CDK also phosphorylates a host of other proteins that are crucial for passage through the cell cycle. There are over 75 targets that have been identified, which control such vital processes as DNA replication and chromosomal segregation. Thus, as the activity of CDK increases and decreases with the levels of cyclin, the various stages of the cell cycle are activated or deactivated.

Although the cyclin–CDK complex is thought to be the central core of the cell cycle oscillator in yeast cells, yeast express multiple cyclins, which appear in successive waves as the cells progress through the cell cycle. Thus, the

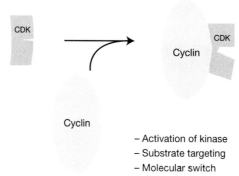

Figure 6.2. Cyclins bind to cyclin-dependent kinases (CDKs), thus acting as a molecular switch that results in the activation of the kinase.

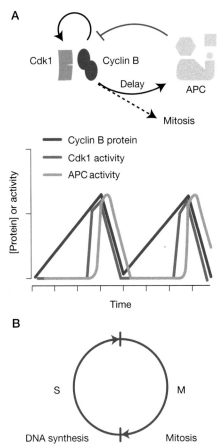

Figure 6.3. Generating simple cell cycle oscillators. (*A*) The cyclin–CDK complex targets specific substrates, including the anaphase-promoting complex (APC). One of the functions of the APC is to target cyclins for destruction. In this way, an oscillator is generated, which is partially responsible for movement from one phase to the next in the cell cycle. (*B*) The period of this negative-feedback oscillator can be very short in embryonic systems, thus aiding the rapid proliferation of cells in a modified cell cycle in which the phase of DNA replication (S phase) is immediately followed by mitosis (M).

models for the oscillator controlling the yeast cell cycle are necessarily more complicated than for early embryonic systems (Fig. 6.4).

The negative-feedback loop between the cyclin that controls the entry into mitosis (**mitotic cyclin**) and APC still appears in contemporary models, as well as a variety of other positive- and negative-feedback loops involving multiple cyclins. Such models predict that cell cycle oscillations are produced by

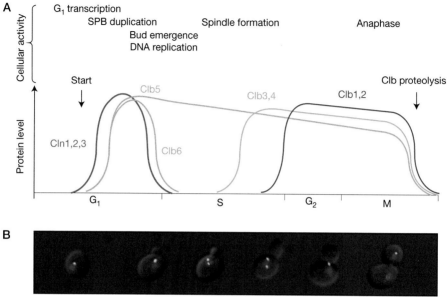

Figure 6.4. In yeast, specific cyclins (Cln1, Clb5, etc.) regulate the entry into different phases of the cell cycle (*A*), beginning with commitment in G_1 phase at a stage termed "Start" and finishing at the end of mitosis with the destruction of B-type (Clb) cyclins. (*B*) Fluorescence images, matched to the cell cycle shown in *A*, showing emergence of a bud and duplication of the SPB (green) during interphase, before separation of the chromatin (blue) in mitosis (M). SPB, spindle pole body.

complex dynamics that include the activation and de-activation of multiple cyclin–CDK complexes (Fig. 6.5).

After the DNA is replicated in S phase, it must be appropriately partitioned so that each daughter cell gets a copy of the DNA. Part of this process involves condensation of the DNA, during which it becomes tightly packed into the form that we normally associate with chromosomes. The replicated **chromatids** (copies of the original chromosomes) are then pulled in opposite directions by the attachment of microtubules to chromosomal regions called **centromeres**. In baker's yeast, one set of chromatids is pulled into the newly formed bud. In contrast, in most animal cells, there is no bud—instead, a gradual tightening occurs through a **contractile ring** made up of actin and myosin, with half the chromatids going into each of the two newly forming cells. In plant cells, a new membrane and cell wall are formed, which divide the mother into two daughter cells. This process of partitioning the DNA and forming new cells is called **cytokinesis.** During the development of plants and animals, a special type of cytokinesis occurs in which the two daughter cells have different fates. One way to

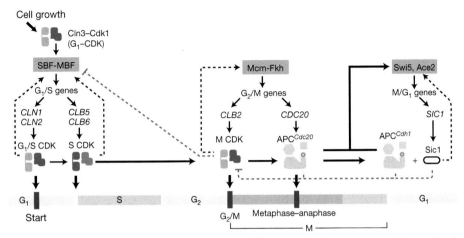

Figure 6.5. Cell cycle oscillations can be regulated by complex dynamics involving multiple cyclin–CDK complexes. One model of cell cycle control involves cyclin–CDK complexes that are part of positive- and negative-feedback loops, which control the timing of the cell cycle. APC, anaphase-promoting complex; CDK, cyclin-dependent kinase. Factors in shaded boxes are key cell cycle regulators.

make an **asymmetric division** is to partition different molecules into the two daughter cells. Although both cells receive the same complement of DNA, they can have different transcription factors loaded into them, which will trigger different expression patterns, ultimately leading to different cell fates.

6.5 Using Reverse Genetics to Understand Cell Division in Animals

In animals, there are some major changes in the cell cycle that occur as the embryo grows and matures to form an adult organism. First, the period of the cell division cycle increases substantially, from minutes to many hours. Second, the mRNA for cyclin is originally synthesized by the mother, but as it is depleted, the embryo must activate transcription from its own genome to express cyclin protein. Finally, although embryonic cell cycles are driven primarily by a single cyclin (cyclin B), **somatic cells** (not egg or sperm) express a variety of cyclins, in a similar fashion to yeast, which are expressed in temporally distinct waves as cells move through successive stages of the cell cycle (Fig. 6.4A).

Just as with yeast, genetic screens have been used in animals to gain insight into the cell division process. Researchers have used the roundworm model system—C. elegans—to look at the early cell division events during embryogenesis.

Instead of treating the worms with a mutagen, such as EMS, investigators took a reverse-genetics approach. In **reverse genetics**, one modifies the activity of specific genes and then observes the effect on the phenotype of interest. This approach became popular once genome sequences were available and new means of interrupting gene activity were developed. In *C. elegans*, a technique called **RNA interference (RNAi)** has proven to be a very effective means of reducing gene expression. A **double-stranded RNA (dsRNA)** is introduced into the worm, and this causes the complementary mRNA to be destroyed. Getting the dsRNA into a worm turns out to be remarkably easy—the worms can be fed on bacteria that express the dsRNA, they can be soaked in a solution of dsRNA or it can be injected into them.

In one such screen, genes that had previously been identified as being expressed in the ovary were systematically targeted with RNAi. As with the Hartwell screen, time-lapse photography under a microscope was used to track the effects on the first set of cell divisions. In nontreated embryos, the nuclei from the sperm and the egg fuse to form the zygote, which then divides to form two cells. These undergo cytokinesis to form four cells, and so on. One of the most striking phenotypes is a complete loss of cytokinesis, whereas the embryo continues to try to effect new cell divisions. It attempts to separate its chromosomes, which instead of being engulfed in new nuclei seem to form small organelles that look somewhat like nuclei. The disrupted gene, in this case, encodes the cytoskeletal element actin, which is known to play an important role in cell division.

6.6 Testing the Cyclin–CDK Oscillator Model

Despite the many changes as cells transit from early embryonic cycles to somatic cycles, cyclin–CDK complexes are thought to remain at the heart of the central cell cycle oscillator. To test rigorously the importance of the cyclin–CDK complex, researchers used reverse genetics to delete the genes encoding all cyclins responsible for passage through S phase and M phase in baker's yeast. If these complexes were at the heart of cell-cycle progression, then oscillations should stop and cell cycle events should arrest. What was observed was that progression through the cell cycle was partially blocked, but some of the oscillatory phenomena seemed to continue. In particular, buds still formed and grew at about the same rate as in wild-type yeast, even though the DNA was not being replicated. Surprisingly, the buds did not pinch off to form new cells, but instead new buds formed on the original bud. In fact, cells lacking cyclin genes can undergo multiple budding cycles in which buds emerge one at a time. Initially, researchers measured the levels of a few RNA transcripts known to be involved in the cell cycle and found that they also continued to oscillate in the absence of cyclins. These oscillations showed the same period as budding cycles, and occurred in the same temporal order as they do in normally cycling cells. These findings were some of the first to suggest that cyclin–CDK complexes might not be the only oscillator controlling periodic cell cycle events.

The observation that the transcript levels of some genes remain periodic in cells lacking S-phase and M-phase cyclins suggested that a transcription-factor-based oscillator might underlie the cyclin–CDK-independent oscillations. If so, then the genes involved should be expressed periodically in cells lacking cyclin genes. To identify these genes, the researchers synchronized cells lacking cyclin genes in early G_1 phase. The cells were then allowed to start the division process, and mRNAs were isolated at different time points. These were fluorescently labeled and hybridized to microarrays.

In normally cycling cells, ~20% of the genome (>1200 genes) is transcribed periodically in discrete cell cycle intervals. The activation of genes at specific times during the cell cycle has been referred to as "just in time" gene expression, and the synthesis of specific proteins at specific times is thought to contribute to the ordering of specific cell cycle events (Fig. 6.6). Surprisingly, in cells lacking S- and M-phase cyclins, ~70% (>900 genes) are expressed on schedule, despite the fact that these mutant cells are unable to replicate DNA or complete mitosis (Fig. 6.6B).

To generate a gene-regulatory network that could be responsible for this oscillatory behavior, the known binding sites for yeast transcription factors were used to form edges between transcription factor nodes and potential target genes whose promoters contain their binding sites. An additional piece of information was the time at which transcript levels peaked for each transcription factor during the cell cycle. Taken together, this allowed a network graph to be constructed for both wild-type and cyclin mutants in which periodically expressed transcription factors were placed on a cell cycle time-line based on the time of peak expression.

In both cases, a network of transcription factors emerged in which expression is serially activated. In these networks, nodes from the end of the cycle are connected to nodes from the beginning, suggesting a means by which the network could function as an oscillator. The network for cells lacking cyclin genes had fewer nodes, probably because fewer genes show periodic behaviors. Interestingly, the smaller network found in the cyclin-mutant cells is a subset of the

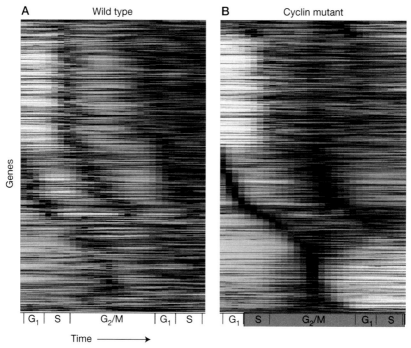

Figure 6.6. Examining the effect of perturbing cyclin expression. In an experiment in which all of the cyclins responsible for passage through S phase and M phase were inactivated in yeast, oscillation of gene expression still occurred, with dynamics similar to those of the wild type. (A) Presentation of the results of microarrays performed at intervals throughout the cell cycle, with each row representing a different gene. In the cyclin mutant (B), the genes are placed in the same order on the y-axis, showing that oscillation is not dramatically impaired in the absence of the cyclins (despite the fact that the mutant cells arrest at the G_1–S border).

larger network observed in normally cycling wild-type cells. The implication is that all of the nodes in the wild-type network are not required for oscillations, but some might play ancillary roles as modulators or effectors of the oscillation.

6.7 Dynamics of Networks Regulating the Cell Cycle

The ability to construct a network graph depicting a cascading network that appears to feed back on itself is not sufficient to show that the network can oscillate. For this, one can use mathematical approaches to model the potential dynamics of the network. One modeling approach is to make **Boolean** connections at each node, using the logic operators AND, OR, etc., to describe the effect of transcription factors on their target genes. A simple, synchronously updating Boolean model was applied to the cell cycle network identified from cells lacking genes encoding cyclins. The dynamics of this model were tested by running thousands of simulations, which showed that the most common outcome or **attractor** was indeed an oscillator, with the only other attractor being a fixed point with all nodes off (Fig. 6.7).

A complementary modeling approach is the use of differential equations. In contrast to Boolean modeling in which each step results in a specific logical choice, with differential equation modeling, time is treated as continuous. The major challenge with this approach is to determine the correct parameters for each differential equation. Usually, these are approximated based on guesses or data from another system. For the cyclin–CDK oscillations of the yeast cell cycle, a model has been generated with more than 50 differential equations using more than 100 parameters. For all models, the real test is how well they make nonintuitive predictions that can be tested by experiments. For the cell cycle, testing of model predictions is an ongoing endeavor.

Figure 6.7. A gene-regulatory network for the yeast cell cycle was generated by combining knowledge of binding sites of transcription factors (in shaded boxes) with the timing of the peak of their expression. A Boolean model based on this network had, as its most common attractor, an oscillator, suggesting that the transcriptional network was sufficient to drive the cell cycle.

6.8 Molecular Networks Underlying Circadian Oscillators

Our body temperature, blood pressure, hormone levels, and a large number of other physiological processes go up and down, with a periodicity of 24 h. In plants, circadian-regulated processes include leaf movements, photosynthesis, and response to pathogens. In both plants and animals, the circadian clock can be reset, or **entrained**, by changes in light–dark cycles, but it will continue for several cycles if individuals are kept in constant light or darkness. The primary role of circadian oscillators is to coordinate cellular and organismal processes with the 24-h light–dark cycle. This allows the organism to anticipate a change in the environment so that it is prepared for events that occur when the sun comes up or goes down.

The circadian clock in plants and animals can function in isolated cells, but frequently it is coordinated across many tissues in the body. In animals, the primary clock is located in a region of the brain called the suprachiasmatic nucleus. Signals from the eyes relay information concerning light–dark cycles, whereas nerves from the suprachiasmatic nucleus direct hormones to other parts of the body.

Although the constituents of the circadian oscillator can vary from organism to organism, most circadian oscillators are based on negative-feedback loops involving transcription factors and the genes that encode them (Fig. 6.8). Circadian oscillations can be driven by a transcription factor network with as few as five genes organized in a negative-feedback loop. In plants, there are two interacting negative-feedback loops, one for the morning and one for the evening, which have been described as resembling the synthetic repressilator circuit (see Chapter 4.13). In animals, there are both transcriptional (affecting gene transcription) and translational (affecting protein synthesis) feedback loops, which include positive loops as well as negative loops.

6.9 A Clock-Like Process for Positioning Lateral Roots

The first evidence for a clock-like process acting in plant roots was periodic expression from a synthetic promoter that responds to the plant hormone auxin. After the wave of expression passed through the root tip, it remained on in a highly localized set of positions along the root (Fig. 6.9A). These turned out to be the sites of lateral root formation (Fig. 6.9B), leading to the localized expression being called **prebranch sites**.

Microarray data revealed periodic expression of other genes in the root. The expression of approximately 2000 genes oscillates in the same phase as the

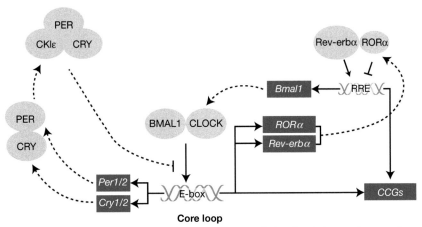

Figure 6.8. Circadian rhythms can be generated through feedback loops that involve transcription factors and signaling molecules. BMAL1, brain and muscle ARNT-like 1; CCG, clock-controlled gene; CK1, casein kinase 1; CRY, cryptochrome; PER, period circadian protein; RORα, retinoid-related orphan receptor-α; RRE, Rev response element.

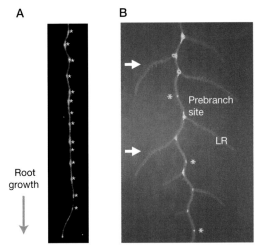

Figure 6.9. DR5 marks prebranch sites. (A) In the model plant *Arabidopsis*, expression of the reporter gene *DR5* was shown to oscillate at the root tip and generate localized expression at periodic intervals along the root. Asterisks (*) indicate prebranch sites. (B) These localized spots of expression were the site of subsequent formation of lateral roots (LRs, arrows).

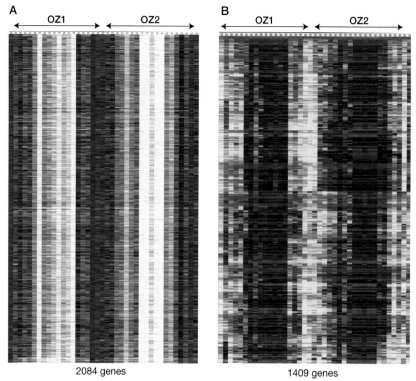

Figure 6.10. Microarray analysis of root tips from *Arabidopsis* seedlings at different stages of the gene oscillation described in Fig. 6.9 identified more than 2000 genes that appeared to oscillate in phase with the original reporter gene (*A*) and more than 1400 genes that appeared to oscillate in antiphase with the original reporter gene (*B*). OZ, oscillation zone.

original synthetic promoter, whereas another approximately 1400 genes appeared to oscillate in antiphase to the expression of the synthetic promoter (Fig. 6.10). Reverse genetics was used to show that some of these genes played important roles in lateral root formation.

These results provided evidence that oscillating expression of genes in the root tip plays a role in determining the position of lateral roots. Currently, it is unknown as to what drives the oscillating gene expression. A somewhat analogous situation exists for the formation of somites in vertebrate animals. In that case, there appears to be an autonomous pacemaker that drives three interlocking feedback loops consisting of both transcriptional and signaling components.

CHAPTER 7

Transcriptional Networks in Development

7.1 What Is Development?

We all know how, with the first cry of a newborn baby, the anxious parents count fingers and toes, and only when the number is verified do they breathe a sigh of relief. It is not surprising that errors can occur in the process that starts with a fertilized egg. This single cell must divide countless times to take on new cellular identities, form organs, and connect tissues with blood vessels and nerves until the baby emerges into the world. This daunting task of programming and execution known as **development** does not end with birth but continues as the child matures and ages. Why do some cells become heart muscle, whereas others become hair follicles? Why will some cells like those of the liver regenerate, whereas other cells such as neurons are incapable of regeneration? What are the properties of the stem cells in the embryo that are able to divide to become any differentiated cell? How does development go awry, and are there ways to repair it when mistakes are made? These are the questions that developmental biology attempts to answer.

7.2 How Do Scientists Study Development?

The very nature of development—the transition from single cell to multicellular organism—is remarkably complex. It is, therefore, not surprising that scientists have looked for ways to reduce the complexity. They have identified organisms that undergo development but are easy to use for genetics. Today, the favored **model organisms** for studying developmental biology include the fruit fly

Drosophila melanogaster, the roundworm *Caenorhabditis elegans*, the zebra-fish *Dario rerio*, the mouse *Mus musculus*, and the plant *Arabidopsis thaliana*.

When you compare an animal embryo with the adult, you find the general form of the body already present in miniature. The basic organization that places the head at one end, the tail at the other, and arms and legs in between is called the **body plan**. One of the most productive genetic screens was performed by Christiane Nüsslein-Volhard and Eric Wieschaus, who searched for mutations that affected the body plan of *Drosophila*. When performing a genetic screen in a **diploid** organism such as *Drosophila*, the effect of the mutations does not show up until sisters and brothers are crossed to obtain homozygous strains. When a mutation is **recessive** (copies on both chromosomes are needed to detect a phenotype) then one-quarter of the progeny will have the phenotype. When a mutation is dominant, such as one that produces too much gene expression, then three-quarters of the progeny will have the phenotype.

The Nüsslein-Volhard screen made a bold assumption—that mutations affecting early stages of embryo development would not arrest embryogenesis at that stage. They performed the screen at a relatively late stage of embryo development and looked for dead embryos with altered body plans. This was greatly facilitated by the fact that the *Drosophila* embryo comprises 14 segments, which can be seen as bands of hair-like projections called **denticle belts**. The screen identified mutations that eliminated some or all of these segments. Of particular note were mutations that removed only the even-numbered segments, called **even-skipped**, or only the odd-numbered segments, called **odd-skipped**.

Many of the genes mutated in this pioneering screen have been identified. Remarkably, most of them turned out to be transcription factors that were expressed in a precise temporal order. The early-expressed genes demarcated broad regions in the embryo, which were then subdivided by later-expressed genes and then further subdivided by other transcription factors to define the 14 segments. Some of the transcription factors were found in a gradient along the embryo, with the highest concentration at one end and the lowest at the other end. There was evidence that different threshold levels of the transcription factor found in the gradient triggered different segmentation events, suggesting that the factor acted as a **morphogen gradient**.

Many of the later-expressed transcription factors, including even-skipped, are **homeobox** transcription factors. What is striking is that their genes are located close to each other on the chromosome and are expressed in the order they are found on the chromosome. Remarkably, when the orthologs were identified in mice and other vertebrates, they were lined up in the same order, expressed in the same relative order during development, and interfering with their expression had developmental consequences, as explained below.

7.3 How Can the Role of Homeobox Genes in Mice Be Studied?

The colinearity of homeobox genes in mice and flies raised the intriguing question as to whether these genes are also involved in segmentation in mammals. The problem is that mammals do not have obvious segments in the same way that insects do. It was also much more difficult to conduct a screen for early embryonic mutations in mice than it was in flies. To answer this question required the development of a technique to knock out the function of specific genes in mice.

To do this, Mario Capecchi and Oliver Smithies combined an ingenious selection scheme with in vitro fertilization techniques. The process by which DNA is exchanged between sister chromosomes is called **homologous recombination**. It requires long stretches of very similar DNA sequence that are recognized by special enzymes, which cut and paste one piece of DNA into the other. The problem is that, when DNA is introduced into a cell, it usually inserts into the genome at a random position rather than undergoing homologous recombination. To increase the likelihood of homologous recombination, the researchers added a gene to the end of the homologous region, which produced a toxic compound when grown in the presence of a drug. If the insertion was random, then the toxic gene would be expressed and the cells would die. By contrast, if homologous recombination took place, the toxic gene would be clipped off and the cells would survive. In addition to using this **negative selection** scheme, the construct also had a **positive selectable marker** to identify only those cells that contained the homologous sequence.

The construct can be introduced into cells isolated from early mouse embryos, and they are subjected to both positive and negative selection. The surviving cells are then reintroduced into an embryo of the same stage and placed in the uterus of a foster mother mouse. The resulting offspring are chimeras derived from both the selected cells and normal cells in the foster embryo. Among the offspring of these chimeric mice are pure-bred selected mice. These heterozygotes are crossed to siblings to produce mice that are homozygous for the knocked-out gene.

In this way, knock-out mice were produced that lacked each of the homeobox genes orthologous to the fly genes. Remarkably, some of these knock-out mice were lacking specific vertebrae and associated structures. This indicated that the molecular mechanism that set up the segmentation pattern in *Drosophila* had been conserved in mammals and was used to determine the identities of different regions along the spine.

7.4 A Circuit Diagram Describing Development in the Sea Urchin

Knocking out genes in mice has been very informative, but it is a costly and time-consuming process. Another model system that has been used to probe the workings of embryogenesis is the sea urchin *Strongylocentrotus purpuratus*. An advantage of the urchin is that the embryonic cell divisions are highly predictable both in time and location. An additional advantage is that cells can be removed and/or moved within the embryo. The ease of manipulation of the sea urchin embryo is one of the major reasons why it was the first to have a comprehensive network generated that describes the major regulatory steps during embryogenesis.

Instead of homologous recombination, the favored approach to reduce gene activity in urchins is to use modified DNA sequences called **morpholino antisense oligonucleotides**. These contain a DNA sequence that is complementary to a specific mRNA, and they are chemically modified to be stable when injected into cells. When they form a complex with the mRNA, they inhibit translation effectively.

Researchers have systematically reduced the activity of genes encoding transcription factors and signaling components at different stages of urchin embryogenesis. They then tested the level of expression of other genes that were expressed at later time points. If they found reduced expression, they inferred that the upstream factor acted as an activator, whereas, if they found increased expression, they inferred that it acted as a repressor. The results of these experiments have been schematized as a circuit diagram (Fig. 7.1).

7.5 Positional Information in a Moving Transcription Factor

In contrast to an animal embryo, if you compare a plant embryo with the mature version of the same plant, you can detect only a rudimentary relationship. This is because plant development is far more dependent on environmental conditions than animal development. Plants have populations of stem cells at either end of the embryo. One set gives rise to all the aboveground organs, and the other set gives rise to the entire root system. Another major difference between

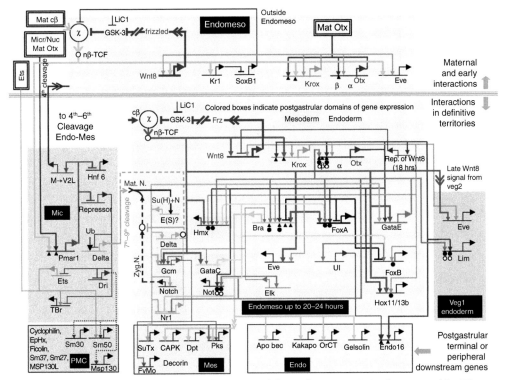

Figure 7.1. A gene-regulatory network generated for embryogenesis in the sea urchin. The transcription factors and signaling elements have been illustrated connected to each other by lines, where arrowheads represent activators and bars signify repressors. The various stages of embryogenesis are denoted by different colored regions of the diagram. Endo, endoderm; Endomeso, endomesoderm; Mat, maternal; Mes, mesoderm; Mic, micromere; N, nucleus; PMC, primary mesenchyme cell; Rep., repression; zyg., zygotic.

plant and animal development is that plant cells are bound by cell walls and thus cannot move. This turns out to be an advantage when studying the process by which cells obtain their identities.

The root of the model plant *Arabidopsis* has a very simple organization, with the different cell types arranged as concentric cylinders (Fig. 7.2A). Another major simplifying aspect of root development is that all the cells in the root are derived from the stem cell population located at the tip of the root (Fig. 7.2B). Under the microscope, the *Arabidopsis* root is translucent, allowing easy visualization of the stem cells and their immediate progeny. In a genetic screen for mutants with shorter roots, two mutants were found where the primary defect was the loss of a cell layer. For both mutants, there was a loss

A   B

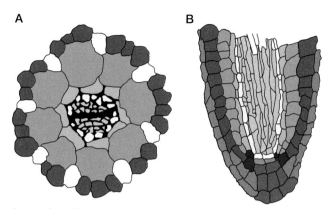

Figure 7.2. Reducing four dimensions to two dimensions. (A) Viewed in cross-section, the root of *Arabidopsis* is formed by concentric cylinders of cells. The four outer layers have radial symmetry, whereas the inner vascular tissue is bilaterally symmetric. (B) Longitudinal section showing the organization of the tissues and the stem cell population (red, epidermis; blue, cortex; green, endodermis; yellow, pericycle; pink, vascular tissue; orange, root cap). The two noncolored cells, which are mitotically inactive, are the center of the stem cell niche and are surrounded by stem cells for each of the tissues.

of the asymmetric division that generates the two internal cell layers—the endodermis and cortex.

The genes for SHORT-ROOT and SCARECROW turned out to encode related members of a plant-specific transcription factor family. SHORT-ROOT has the interesting property that it can move from cell to cell through pores called **plasmodesmata**. Thus, the SHORT-ROOT protein acts both as a transcriptional regulator and provides positional information. Its directional movement from the vascular tissue is crucial in generating the asymmetric division and specifies endodermal identity for the tissue adjacent to the vascular cylinder. To understand how SHORT-ROOT controls the asymmetric division of the stem cell, an inducible version was constructed and transformed into the *short-root* mutant. A time-course was run, with mRNA isolated every 3–6 h from cells in the mutant layer, which were marked with a fluorescent protein and captured using cell sorting. Among the transcripts that peaked when asymmetric divisions were occurring was a D-type cyclin. Further analysis showed that this cyclin was specifically expressed in the stem cells just prior to the asymmetric division and its promoter was bound by both SHORT-ROOT and SCARECROW. Thus, a direct link was identified between development regulators and cell cycle components involved in a key asymmetric cell division.

CHAPTER 8

Understanding Complex Traits

8.1 What Is a Complex Trait?

When we think of inherited traits, we know that some, such as albinism, are caused by mutations in single genes. There are also diseases such as sickle cell disease that are caused by alterations in single genes. However, for the vast majority of physical traits and diseases, multiple genes come into play. Examples include hypertension, cardiovascular disease, diabetes, epilepsy, and autism. These **complex traits** affect between 1% and 23% of the American population, whereas **monogenic traits** generally have frequencies below 0.1%. For most complex traits, the severity of the symptoms depends greatly on the environment. Identical twins with the same genome growing up in different environments can have very different disease profiles.

In agriculture, most desirable traits, such as drought resistance, are also governed by multiple genes. This complicates immensely efforts to breed crops with improved performance. Identifying the genes involved in complex traits and understanding how each one contributes is a daunting task. Applications of genome-wide approaches and systems-biology analysis tools hold the promise of developing more effective disease treatments and improved crops.

8.2 How Do You Know If a Complex Trait Is Heritable?

Traits that are controlled by single genes, also called **Mendelian** traits, tend to be all or nothing. If a child receives both of the albino genes, then she will have the albino trait, but if she receives only one, she will appear normal.

For complex traits, such as height, there is a continuous distribution. Line up all of the people in a classroom according to height and the distribution resembles a **bell curve** (called a **normal distribution**). When a trait is characterized by a continuous distribution of the phenotype and the phenotype can be quantified, it is called a **quantitative trait**. The more genes involved in a complex trait, the more likely the phenotype will follow a normal distribution.

To determine whether a complex trait is heritable, a favored approach is to compare the incidence in identical twins with that in nonidentical twins. An example is epilepsy in which there is concordance between identical twins of 62%, whereas in nonidentical twins it is only 18%. This gives a direct measure of heritability, which in this case is considered to be 62%.

8.3 How Do Genes Interact to Produce a Complex Trait?

There are very few cases where it is well understood how genes interact to produce a complex trait. For most complex disease states, this is very much a work in progress. An informative example can be found in plants in the interaction of a small number of genes to form a flower.

In *Arabidopsis*, the flower comprises four organs. On the outside of the flower are leaf-like organs called **sepals** (Fig. 8.1). Within the sepals are the **petals**, which surround the male sexual organs, the pollen-carrying **stamens**. In the center of the flower is the female sexual organ consisting of fused **carpels** (Fig. 8.1). When viewed from the top, the organs appear to be arranged in

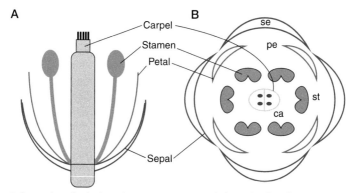

Figure 8.1. Schematics depicting the organization of the whorls of an *Arabidopsis* flower shown in a cutaway (*A*) or from the top (*B*), with the four whorls arranged from outer to inner: sepal (se), petal (pe), stamen (st), and carpel (ca).

Figure 8.2. Mutations in the *agamous* gene result in sepals in whorl 1, petals in whorl 2, petals in whorl 3, and sepals in whorl 4. The two flowers are from two different alleles exhibiting somewhat different floral morphologies.

whorls, with sepals in the outer whorl, which will be referred to as whorl 1, petals in whorl 2, stamens in whorl 3, and carpels in whorl 4 (Fig. 8.1).

Genetic screens identified three types of mutations that affected floral organ formation. A mutant named **agamous** (**ag**) has sepals in whorl 1, petals in whorl 2, petals in whorl 3, and sepals in whorl 4 (Fig. 8.2). In other words, the stamens have been transformed into petals and the carpels have been transformed into sepals. An additional defect is that the flower appears to repeat itself (Fig. 8.2). This is known as **indeterminate** growth. A second mutant, **apetala2** (**ap2**), has no sepals or petals. Instead, it has carpel-like tissue in whorl 1, stamens in whorl 2, more stamens in whorl 3, and carpels in whorl 4 (Fig. 8.3). In this case, the sepals have been transformed into carpels and the petals have been transformed into stamens. The third mutant, **apetala3** (**ap3**), has sepals in whorl 1, sepals in whorl 2, carpels in whorl 3, and carpels in whorl 4 (Fig. 8.4). For this mutant, the petals have been transformed into sepals and the stamens have been transformed into carpels. There is a second mutation, **pistillata** (**pi**), that has a phenotype identical to that of *ap3*. I will refer

Figure 8.3. Mutations in the *apetala2* gene result in carpel-like tissue in whorl 1, stamens in whorl 2, more stamens in whorl 3, and carpels in whorl 4.

Figure 8.4. Mutations in the *apetala3* gene result in sepals in whorl 1, sepals in whorl 2, carpels in whorl 3, and carpels in whorl 4.

to these last two mutations together as *ap3/pi*. To be able to compare the mutants with the wild type, it is useful to make a table with the organs listed in each whorl (Table 8.1).

As mentioned earlier, a fundamental principle of genetics is that, where there is evidence of a difference between mutant and wild type, this indicates the function of the normal gene product (i.e., the encoded protein). Looking at the table, we can see that the normal AGAMOUS gene function must be active in whorls 3 and 4 as it is in those positions that there are differences with wild type. There is the additional difference of the indeterminate flower, the flower within a flower, which appears to be regulated by the same gene. By the same reasoning, the normal function of APETALA2 must be in whorls 1 and 2, whereas the normal function of APETALA3/PISTILLATA must be in whorls 2 and 3. If we bring together this information, we see the beginnings of a model for how the three genes interact to form a flower (Fig. 8.5). We can hypothesize that, to form sepals, we only need the gene encoding AP2. To form petals, we need the interaction of AP2 and AP3/PI. To form stamens, we need the interaction of AP3/PI and AG, whereas to form carpels we only need AG. This model is known as the ABC model of flower development.

Table 8.1. Phenotype of flower structures of wild-type and mutant *Arabidopsis*

Genotype	Whorls			
	1	2	3	4
Wild type	Sepal	Petal	Stamen	Carpel
agamous	Sepal	Petal	Petal	Sepal[a]
apetala2	Carpel	Stamen	Stamen	Carpel
apetala3/pistillata	Sepal	Sepal	Carpel	Carpel

[a]Denotes an indeterminate flower—that is, a flower within a flower within a flower, etc.

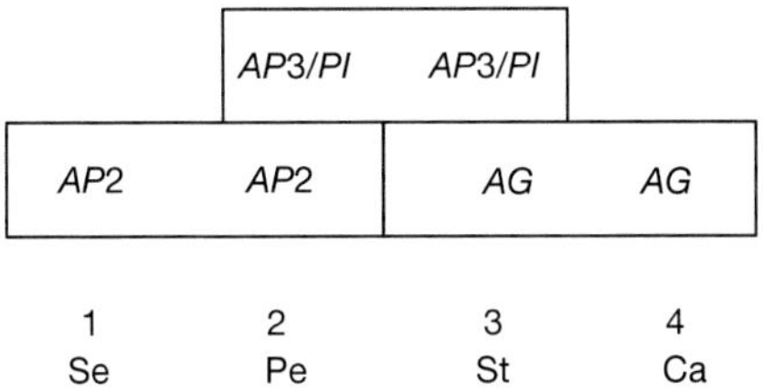

	AP3/PI	AP3/PI		
AP2	AP2	AG	AG	

1	2	3	4
Se	Pe	St	Ca

Figure 8.5. Model depicting the interactions of the different genes and how they are required for the four whorls. *AP2, APETALA2*; *AG, AGAMOUS*; *AP3/PI, APETALA3/PISTILLATA*; Ca, carpel; Pe, petal; Se, sepal; St, stamen.

8.4 Testing a Multigene Model

To test this model, we can make combinations of mutations. It is a useful exercise to see whether you can predict, based on the ABC model, what the phenotype would be of the following combinations: *ag* and *ap3/pi*, *ap2* and *ap3/pi*, and *ap2* and *ag*. While working out the consequences of combining these mutations, it should have become apparent that there was a key piece of information needed for the model. When AG is mutated, what happens to the activity of AP2 and vice versa? From the table, we can see that when AG is mutated then petals are formed in whorl 3 and sepals in whorl 4. According to our hypothesis, to form petals we need the combined activity of AP2 and AP3/PI, and to form sepals, we need the activity of AP2. Therefore, the *ag* phenotype indicates that AP2 activity must now be present in whorls 3 and 4. The conclusion is that, in wild type, AG prevents AP2 from acting in whorls 3 and 4 (Fig. 8.5). Similar logic holds for AP2 preventing AG from acting in whorls 1 and 2. With this refinement of the model, we can now predict that the double mutant, *ag,ap3/pi* should have sepals in all four whorls, as this would leave only AP2 activity in all four whorls. When the double mutant was made, it matched the prediction, with the additional feature of being indeterminate owing to the lack of AG activity (Fig. 8.6). The model would also

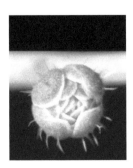

Figure 8.6. A double mutant of *agamous* and *apetala3* results in sepals in all four whorls.

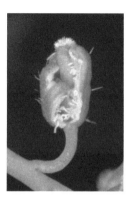

Figure 8.7. A double mutant of *apetala2* and *apetala3* results in carpel tissue in all four whorls.

predict that the double mutant *ap2,ap3/pi* would have carpels in all four whorls as only AG activity would remain. This was indeed what was found (Fig. 8.7). More difficult to predict was the phenotype of the *ap2,ag* double mutant. We do not have any basis for predicting the phenotype when AP3/PI is the only activity in a whorl, as would be the case for whorls 2 and 3, or when none of the three genes was active, as would be the case in whorls 1 and 4. It turns out that the organs in whorls 2 and 3 have features of both stamens and petals (Fig. 8.8). In whorls 1 and 4, the organs resemble leaves.

With this information, you should not be surprised to learn that the phenotype of the triple mutant *ag,ap3/pi,ap2* is an indeterminate flower with leaf-like organs in all whorls (Fig. 8.9). This highlights the remarkable interaction of just four genes to form all of the organs in a flower. Moreover, it suggests that flowers might have originated from leaves, something that the German philosopher Goethe proposed before the theory of evolution was formulated. It should be pointed out that, in this example, each of the mutations results in a complete loss of gene activity. In nature, the more common situation is that the genes have

Figure 8.8. A double mutant of *apetala2* and *agamous* results in whorls 2 and 3 having features of both stamens and petals, whereas whorls 1 and 4 resemble leaves.

Figure 8.9. A triple mutant of *apetala2*, *agamous*, and *apetala3* results in leaf-like organs in all four whorls.

differing levels of activity, resulting in combinations of different alleles. There is increasing evidence that much of the diversity in flowers found in nature is due to different alleles of the orthologs of these same four genes.

When thinking about how these floral organ genes interact, it is useful to place them in the context of two basic types of genetic interactions: **additive** and **epistatic**. As its name suggests, additive interactions are when the phenotype resulting from the combination of mutations is the addition of the phenotypes found when the mutations are alone. Epistatic interactions occur whenever the interaction is nonadditive. One example of an epistatic interaction is when a double mutant has essentially the same phenotype as only one of the two single mutants. This would be the case for mutations in the enzymes of a metabolic pathway, such as the set of enzymes that generate ethanol in yeast (see Chapter 4.4). If a mutation in a gene encoding an upstream enzyme in the pathway is combined with a mutation in a gene encoding a downstream enzyme, then the double-mutant phenotype will resemble that of the upstream mutant. Another type of epistatic interaction is when the double-mutant phenotype does not resemble either one of the single-mutant phenotypes. These are called synergistic interactions and apply to all of the floral organ genetic interactions. For complex traits that involve many genes, both additive and epistatic interactions occur. The fact that epistatic interactions frequently occur complicates greatly the job of predicting the effects of genetic interactions in complex traits.

8.5 What Is the Role of Environmental Factors in Complex Disease Traits?

As mentioned above, another complicating factor in understanding complex traits is the environment. A striking example is adult-onset diabetes, the frequency of which has increased dramatically in the American population over the past 50 years. There is little evidence of major changes in the genetic makeup of the population over the same time period—thus, the likely culprit is lifestyle changes such as the development of obesity and lack of exercise.

Strong evidence in support of the role of the environment in diabetes comes from a study of the Pima Indian tribe. The study compared tribal members living in Mexico with those living in Arizona. There is a clear genetic disposition for diabetes as over 50% of the Arizona tribe who are over the age of 35 have the disease. However, the tribal members living in Mexico, with presumably similar genetic makeup, have rates of diabetes that are no higher than those of the general population. The striking difference between the two groups is their lifestyles. In Arizona, the tribal members eat a "normal" high-fat American diet and live a sedentary lifestyle (<2 h of hard labor per week), whereas their Mexican counterparts eat a low-fat diet and average 23 h per week of hard labor.

8.6 How Are the Genes Underlying Complex Traits Mapped?

Clearly, the influence of environment on complex traits makes identification of the underlying genes challenging. There are two main approaches used to identify genes responsible for complex traits. Both look for correlations between the trait and specific regions of DNA. The higher the correlation, the more likely there is a gene of interest in that region of DNA. To understand how this works, we must introduce the concept of **linkage**. Unlike all other cells in the human body, sperm and egg cells only have one copy of each chromosome. The process that results in generation of a **haploid** genome (as opposed to a **diploid** genome, with two copies of each chromosome) is called **meiosis**. In addition to reducing the number of chromosomes, another important feature of meiosis is **recombination**. This is the process by which the DNA from each parent is mixed together to form new chromosomes that contain regions from both parents. During **homologous recombination**, the chromosomes from each parent align and special enzymes cut and swap portions of the DNA. It is intuitively obvious that genes that are close to each other on a chromosome are likely to remain so when recombination occurs, whereas genes that are far apart are likely to be shuffled. Linkage is the measure of how close genes are together based on how often they are shuffled during recombination, with the unit being the **centiMorgan (cM)**, named in honor of Thomas Hunt Morgan, one of the pioneers of modern genetics.

To determine linkage relationships, there must be differences between the maternal and paternal chromosomes. These can be differences in the way a gene acts: for example, one allele causes white-eye color whereas the other allele causes the eyes to be red. It can also be slight differences such as a single nucleotide in the DNA sequence called a **single-nucleotide polymorphism (SNP)**, which have no consequence for the phenotype. In both cases, these are referred to as **genetic markers** and can be used to generate a **genetic map** that orders the markers and gives relative distances between them based on recombination frequencies.

To map a quantitative trait, researchers look for linkage between the trait and markers either in a family or across a large population. For many years, family-based linkage was preferred as there were fewer variables. The main drawback was that the number of individuals with the trait was usually small, thus limiting the ability to narrow down the DNA region responsible for the trait. In the past decade, linkage through **genome-wide association studies (GWASs)** has become the more commonly used approach to identify DNA regions responsible for complex traits. These chromosomal regions are called **quantitative trait loci (QTLs)**. Conceptually, it would seem straightforward to look for linkage between a trait—for example type II diabetes—and DNA markers among people with the disease and a lack of linkage for people who do not have the disease. One of the primary complications, as we have seen, is the influence of the environment, so that individuals with genes that predispose them to a disease such as diabetes do not manifest the disease because of what they eat or the type of work they do. There are other complicating factors such as how much recombination has actually taken place in a population, known as **population structure**, and the frequency of the alleles that predispose to the trait. In general, the larger the number of individuals analyzed and the more diverse the population, the more likely to narrow down the QTLs.

8.7 How Are Genes Identified That Predispose People to a Disease?

Even when strong QTLs are identified, the region of the DNA encompassed by each QTL can contain hundreds of genes. We will look at an example in which two complementary methods were used to identify a gene directly involved in **Crohn's disease**, a debilitating affliction that affects approximately 500,000 people in North America. Symptoms include vomiting, diarrhea, and bleeding of the intestine, with associated inflammation of the digestive tract. Evidence that Crohn's disease has a genetic component came from the fact that about

20% of affected individuals have at least one relative with the disease. Family linkage studies identified several QTLs involved in the disease, with one of particular interest on chromosome 16.

Two groups ended up converging on the same gene on chromosome 16. After getting an approximate map position, one group looked for additional markers that were linked to the phenotype in multiple families. They painstakingly homed in on markers that were more and more tightly linked to the disease. The second group used a **candidate gene** approach. They scanned the genes in this region and guessed that the *NOD2* gene might play a role based on evidence that the NOD2 protein is involved in detection of bacteria in plants. To confirm their guess, they sequenced the gene in a large number of Crohn's patients and found several different mutations that would affect gene function. One example was insertion of a nucleotide that would cause a frameshift mutation producing a shortened version of the NOD2 protein. Using PCR, they showed that, in a family, the mutated form of the gene was transmitted from parents to affected siblings. These researchers went on to show that the mutated version of NOD2 produced an altered response of the innate immune system. This is in keeping with current views of Crohn's disease as involving a deficient response of the immune system to bacteria found in the digestive tract.

8.8 How Can Systems Biology Contribute to the Analysis of Complex Traits?

In the case of Crohn's disease, the identification of a single gene provided insights that could be used in designing new therapies. Unfortunately, this has been the rare exception to date. All too often, the genes identified as underlying QTLs for complex diseases have been of relatively small effect, and the potential for new therapeutic approaches has been limited. One possibility is that, in GWASs, only alleles found in a relatively large number of individuals are tracked and that the alleles that cause complex traits are relatively rare. Another view is that most studies have not yet been broad enough to identify the full range of disease-related alleles. One clear success has been identifying alleles that affect response to certain medications. An example is the identification of alleles that determine the response to warfarin, a commonly used blood thinner. Determining the correct dosage of warfarin is crucial as too little can result in blood clots, whereas too much can cause internal bleeding. It is hoped that knowledge of which alleles are present in a patient will lead to more accurate dosage.

It has been suggested that systems biology holds the key to a better under-standing of how multiple genes interact to generate a complex trait. Although there are few good examples to date, the reasoning is that the genes that contribute to a complex trait are likely to form networks. These could be tran-scriptional, signaling, or metabolic networks, or a combination of these. Once the network architecture is understood, then the challenge would be to identify key regulators or bottlenecks in the network and design therapies to modulate them. Almost certainly, it will not be sufficient to identify only the network, but it will be essential to understand its dynamics as well. This is potentially one of the most important contributions that systems biology can make.

Networks in Human Evolution

9.1 How Similar Are We to Apes?

When comparing ourselves with our closest relative—the chimpanzee—we tend to see the differences—the amount of hair, the ability to talk, the way we walk, etc. Yet, at the level of our DNA, there is only about a 5% difference. The human lineage branched off from our great ape cousins approximately five to seven million years ago, which is relatively recently on an evolutionary time-scale. There are other important traits that differ between humans and apes, including behaviors such as social organization, as well as susceptibilities to different diseases. These phenotypic changes happened gradually in ancestral humans and in our extinct relatives from the evolutionary branch of the human lineage. As they diverged from apes, they expanded into new environments, such as from forested areas into open grasslands, which, in turn, caused a change in diet. Later, new behaviors led to adaptations such as the development of agriculture, further altering metabolic processes. New environments also exposed them to new pathogens, to which their bodies learned to respond. Many of these changes left traces in the sequence of the human genome. What are the changes underlying uniquely human traits? Are humans still evolving?

9.2 What Are the Genetic Differences between Humans and Chimpanzees?

At a coarse scale, there are only a few differences between the genomes of human and chimpanzee. The most obvious is that two ancestral chromosomes have fused in modern humans to form what is now defined as our chromosome 2; therefore humans have a diploid set of 46, not 48, chromosomes. There are a handful of other major rearrangements, in which large sections of a chromosome have inverted, so that the genes along the chromosome are no longer in the same linear order.

One way to look for regions important to human traits is to take an evolutionary approach to see which regions of the human genome have changed recently—that is, are specific to human evolutionary history. Comparisons between human and other primate genomes, in which the time since divergence is relatively short, can provide information about recent functional changes, especially in noncoding regions. By aligning the genomes of humans and their close relatives (Fig. 9.1), we can identify regions that are either novel within humans or that have changed in the human or primate lineage. These, then, are the regions that can be functionally examined to see whether they might have played a role in the evolution of specific human traits.

Evidence for **negative selection** is when there is little change in the protein sequence over evolutionary time, with the inference that any change will have a deleterious effect on the organism. In contrast, sequences that have been under positive selection can be identified by comparing their rate of change as compared with other genomic sequences that are unlikely to be under selection. This can be in the form of a **selective sweep**, in which a gene and neighboring sequence show few changes when compared across a population. Selective sweeps occur when a mutated gene confers an evolutionary advantage to the individual carrying that gene. The progeny of these individuals will be more numerous and rapidly take over the population. Studies have found evidence

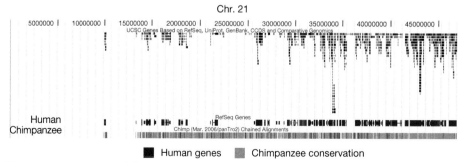

Figure 9.1. Alignment of the gene sequences of human chromosome 21 with those of the homologous chromosome of the chimpanzee *Pan troglodytes*, showing the extensive level of conservation.

for positive selection in humans for genes related to sensory perception, such as olfaction, as well as in genes related to the immune system. In the case of immune-related genes, changes to the amino acid sequence in these genes might have been adaptive, as humans were exposed to new pathogens when they migrated into new environments.

These scans for selection are particularly useful in examining noncoding, possibly regulatory, regions of the genome. Here, potential signals of change are contrasted with other noncoding regions presumed to be under **neutral selective pressure**, such as introns. Neutral regions should be accumulating mutations at a background rate, neither constrained nor accelerated by selective forces. An unusually fast or slow rate of change in the regulatory DNA could be evidence for selection on the control of a gene or set of genes. In the human noncoding regions analyzed, a signal of positive selection is seen in regions associated with genes involved in neural functioning and metabolism (Table 9.1). An interpretation of both the coding and noncoding patterns might be that genes involved in immune functions need to change their protein structure to keep up with the changing structure of the pathogens they are responding to. In contrast, genes involved in processes such as neural development and metabolism are controlled by noncoding changes that alter the amount, timing, or location of gene expression.

Table 9.1. Positive selection associated with genes involved in neural functioning and metabolism

Category	Number of genes	P value on:	
		Human branch	Chimpanzee branch
Protein folding	70	0.0067^a	0.77
Other neuronal activity	31	0.013^b	0.039^c
Neurogenesis	133	0.013^b	0.032^c
Glycolysis	21	0.014^b	0.72
Carbohydrate metabolism	210	0.020^b	0.017^b
Ectoderm development	169	0.020^b	0.11
Mesoderm development	161	0.024^b	0.17
Synaptic transmission	25	0.025^b	0.34
Vision	64	0.025^b	0.15
Oncogene	25	0.045^c	0.46
Anion transport	31	0.049^c	0.17

[a] $P < 0.01$.
[b] $P < 0.03$.
[c] $P < 0.05$.

Studies comparing the genome-wide expression levels between humans and chimpanzees have noted that there are some differences in the relative levels of gene expression in different tissues between these two species. One study examined the difference in gene expression levels between chimpanzees and humans in the brain, liver, heart, kidney, and testis. For most genes, there was a pattern consistent with neutral evolution, in which most expression differences are selectively neutral and might have no phenotypic effect. The strongest signal of positive selection on gene expression was in the testes within both the human and chimpanzee lineages, as compared with other tissues. However, there was also evidence of positive selection on protein-coding regions in testes-expressed genes, illustrating a general pattern of positive selection on the male reproductive system in both lineages.

One would expect that the phenotypic and behavioral differences between humans and chimpanzees, many of which appear to be a result of differences in brain structure and function, would be attributable to striking differences in gene regulation between the brains of humans and chimpanzees. However, there is less change in expression in brain-expressed genes than in other tissues, such as the liver or testes, possibly because of constraints on the complicated expression patterns of many brain-expressed genes. Further studies will need to determine whether the genes with altered expression in specific tissues show evidence of natural selection in noncoding regions.

9.3 What Are the Genomic Differences among Humans?

When comparing two human genomes, the most common variants are SNPs, with the average difference of 1 per 300 bp. Small repeat sequences also play an important role in human genetic variation. Approximately 45% of the human genome comprises repeats called long interspersed nuclear elements (LINEs) or short interspersed nuclear elements (SINEs). These two length classes encompass many different types of repetitive elements. One important example is the *Alu* **retrotransposons**. Sometimes, *Alu* elements insert within a coding or important control region of a gene, causing improper transcript or protein expression, which can lead to disease. *Alu* insertions account for ~0.1% of all human genetic disorders, including hemophilia and various forms of cancer.

Other sources of variation are insertion or deletion of longer stretches of DNA sequence, or rearrangements of large regions of the chromosomes. Both of these can add to **copy number variation** (**CNV**) of a stretch of DNA. The size of CNVs can be anywhere from 1000 bp to several million base pairs. A CNV can encompass a single gene or a region with multiple genes. More copies of a gene, or a change in its genomic location, can dramatically alter the expression levels of the gene, which can lead to disease. A number of recent studies have found large impacts of CNV in susceptibilities to diseases such as schizophrenia and cancer.

To catalog some of the common variation within human populations, the **HapMap Project** has identified SNPs from a large number of related individuals in populations of Asian, African, and European ancestry. Genetic and fossil evidence argues that modern humans were present in a small population in Africa around 150,000 years ago. Most of the genetic diversity that exists in humans comes from this population, and so is shared between all modern populations. A small population size also means that fewer recombination events have taken place, so not much recombination has occurred between SNPs near each other on the same chromosome. Because of this, humans have extended **haplotypes**. Haplotypes are long stretches of base pairs along the chromosome, which are transmitted together during meiosis. Haplotypes are only broken up at areas of recombination between the chromosomes. SNPs within haplotype blocks can be used as proxies, or tags, of differences between the entire blocks of sequence. Tag SNPs can then be genotyped for a large number of individuals at much less expense than whole-genome sequencing. Approximately three million of the polymorphisms common to multiple populations have been mapped by the HapMap Project. This information can be used to explore health and disease risks for different individuals or groups, as well as how individuals respond differently to medications or environmental factors.

Next-generation sequencing is being used to look at genotype in a more detailed way. The **1000 Genomes Project** is acquiring full genomic sequences of more than 1000 individuals from all over the world. These sequences will provide an unprecedented catalog of human genetic variation. This depth of sequencing data from multiple populations will provide much greater resolution as to the variation in genomic sequence, copy number variation, and genomic rearrangements that occur in human populations.

Beyond simple genome sequence, there are also a number of initiatives to map variation in higher-order chromosomal modifications between humans, and between tissues. The **ENCyclopedia Of DNA Elements (ENCODE)** and **Human Epigenome Project** are examples of projects using next-generation sequencing technologies to map regulatory information and chromatin modifications, including DNA methylation and histone modifications.

9.4 Are Humans Still Evolving?

After genes with potentially important consequences to human health and disease have been identified using genome-wide approaches, specific genes and sequence variants can be examined in more detail. Some polymorphisms within and between human populations have clear effects on human pigmentation, specifically in hair, eye, and skin color. Many of these variants lie within the melanocortin gene pathway. The best-studied gene in this pathway is the gene encoding the **melanocortin 1 receptor (MC1R)**, which makes a membrane-bound protein that receives external signals on the surface of pigment-producing melanocyte cells. Coding changes in *MC1R* are responsible for changes in many forms of human pigmentation, as well as variations in coat and skin colors in many other animals, such as horses, dogs, cattle, and pigs. This gene is highly polymorphic within humans. Beyond *MC1R*, a number of other genes in the pigmentation pathway are associated with changes in pigmentation, some affecting different tissues and creating different color variants—for example, fair, red, or dark hair color.

One of the most striking examples of continuing human adaptation is in the evolution of adult tolerance to dairy products in specific populations. This phenotype has evolved independently multiple times in recent human history (<10,000 yr) in populations with a history of animal domestication for the consumption of milk. Lactose tolerance, or more correctly lactase persistence, is defined as the ability to digest milk products after weaning and into adulthood. The genetic basis for this trait is the continued expression of **lactase-phlorizin hydrolase** encoded in the *LCT* gene, which is normally switched off after weaning. The protein made by the *LCT* gene is essential for the digestion of milk sugars. In all of these lactase-persistent populations, there is a signature of strong positive selection around the *LCT* gene. Interestingly, the regulatory mutations that allow for continued *LCT* transcription into adulthood are not near the start site of *LCT* transcription but are located 14 kilobases upstream, within an intron of the neighboring *MCM6* gene. The genetic changes that alter the regulation of *LCT* are also not at the same location in different populations, showing that parallel genetic changes can lead to the same phenotypic trait. In geographically disparate human populations, mutations within the same intron are involved in the lactase persistence phenotype, although these mutations are at different nucleotide positions. The different SNPs also statistically explain different amounts of this phenotype in each of the populations. This example highlights the complex regulation that must be understood in linking changes in transcription to their phenotypic consequences.

Human evolution has also been shaped by infectious pathogens, such as the protozoan parasites that cause malaria or viruses such as HIV. There are at least three genes that appear to have evolved in specific populations in areas with endemic malaria. The evolution of malaria resistance in specific human populations shows how regulatory regions might have been important during recent human evolution. The gene **Duffy antigen/chemokine receptor** (*DARC*) is a **chemokine receptor** that is normally bound by the small signaling molecules—chemokines—used by the immune system, but when expressed on the membrane of red blood cells, it provides an entry point for the malarial parasite *Plasmodium vivax*. The causal mutation under selection disrupts a binding site for the erythroid-cell-specific GATA1 transcription factor within the *cis*-regulatory region of DARC. This causes a loss of the receptor specifically on red blood cells, thus providing resistance to *P. vivax*, but does not disrupt *DARC* expression in other tissues (as they do not express GATA1). This highlights the roles regulatory sequences can play in the subtle modulation of expression, not just in levels, but also in location.

There is also evidence of selection acting on genes involved in behavior and cognition. For example, levels of the neuropeptide products of the **proenkephalin-B (prodynorphin) (*PDYN*)** gene are associated with schizophrenia and temporal lobe epilepsy and are specifically associated with a 68-bp repeat in the *cis*-regulatory region, which shows a signature of positive selection during recent human evolution. This implies a beneficial *cis*-regulatory change that could affect gene expression, which occurred during our divergence from the other great apes. There is also evidence for **balancing selection** in modern populations, indicating the benefit of maintaining multiple haplotypes within the human population. The case of *PDYN* might provide an example of the less common, but essential, selection operating at the level of gene expression during the evolution of the human brain.

CHAPTER 10

Quantitative Approaches in Biology

A T THE BEGINNING OF MY RESEARCH SEMINARS, I frequently show a short video of a flock of birds wheeling through the sky as it tries to ward off a predator. I use it as an analogy for emergent properties—the flock forms and acts through the connections between individual birds. However, it can also be used to illustrate another key aspect of biology—it is a complex, dynamic system undergoing rapid changes over time. As you have, hopefully, come to appreciate from this book, the networks acting in a cell are far more complex than a bird flock in that they are made up of thousands of different components whose interactions change over time. It is the difficulty of dealing with this dynamic complexity that has brought into biology an increasing number of scientists trained in mathematics, computer science, statistics, physics, and engineering.

Biologists in certain subfields, including quantitative genetics and biophysics, have long used sophisticated mathematical approaches. However, much of biology, and particularly molecular biology, rarely required quantitative analysis. What changed the field were large-scale experimental platforms such as microarrays. The data generated were so massive that additional experiments beyond a certain number of replicates did not help. Statistical analyses were the only solution.

As biologists began to grapple with molecular networks that changed over time, they also discovered the need for mathematical-modeling approaches that would capture the dynamics and allow them to predict outcomes. Differential-equation-based modeling is an obvious place to start, but a crucial problem is that it requires a large number of parameters, most of which are unknown. Approaches such as Bayesian and Boolean modeling can be attractive alternatives in that they do not require many biological parameters. However, they are limited in other ways.

What is increasingly clear is that mathematical approaches invented to deal with physical phenomena might not work well for biological problems. This would argue for a new quantitative field with its roots in biology and a set of methods that embrace the complexity and dynamic nature of biological phenomena. With a new mathematics aimed at solving biological problems, biology could finally become a truly quantitative discipline.

Suggested Further Reading

Textbooks

Alberts B, Johnson A, Lewis J, Raff M, Roberts K, Walter P. 2008. *Molecular biology of the cell*, 5th ed. Garland Science, New York.

Alon U. 2006. *An introduction to systems biology: Design principles of biological circuits.* Taylor & Francis Group/Chapman & Hall/CRC, Boca Raton, FL.

Brown SM, ed. 2013. *Next-generation DNA sequencing informatics.* Cold Spring Harbor Laboratory Press, Cold Spring Harbor, NY.

Campbell NA, Reece JB, Urry LA, Cain ML, Wasserman SA, Minorsky PV, Jackson RB. 2008. *Biology*, 8th ed. Pearson, Boston.

Griffiths JF, Wessler SR, Carroll SB, Doebley J. 2012. *Introduction to genetic analysis*, 10th ed. WH Freeman, New York.

Hartwell L, Hood L, Goldberg M, Reynolds A, Silver L. 2011. *Genetics: From genes to genomes*, 4th ed. McGraw-Hill, New York.

Lodish H, Berk A, Kaiser CA, Krieger M. 2012. *Molecular cell biology*, 7th ed. WH Freeman, New York.

Watson JD, Baker TA, Bell SP, Gann A, Levine M, Losick R. 2014. *Molecular biology of the gene*, 7th ed. Pearson, Boston.

Historical Perspective of Molecular Biology

Judson HF. 1996. *The eighth day of creation: Makers of the revolution in biology* (commemorative edition). Cold Spring Harbor Laboratory Press, Cold Spring Harbor, NY.

Primary Literature

Allison AC. 2009. Genetic control of resistance to human malaria. *Curr Opin Immunol* **21:** 499–505.

Coen ES, Meyerowitz EM. 1991. The war of the whorls: Genetic interactions controlling flower development. *Nature* **353:** 31–37.

Davidson EH, Rast JP, Oliveri P, Ransick A, Calestani C, Yuh CH, Minokawa T, Amore G, Hinman V, Arenas-Mena C, et al. 2002. A genomic regulatory network for development. *Science* **295:** 1669–1678.

Hogenesch JB, Ueda HR. 2011. Understanding systems-level properties: Timely stories from the study of clocks. *Nat Rev Genet* **12:** 407–416.

Moreno-Risueno MA, Van Norman JM, Moreno A, Zhang J, Ahnert SE, Benfey PN. 2010. Oscillating gene expression determines competence for periodic *Arabidopsis* root branching. *Science* **329:** 1306–1311.

Orlando DA, Lin CY, Bernard A, Wang JY, Socolar JE, Iversen ES, Hartemink AJ, Haase SB. 2008. Global control of cell-cycle transcription by coupled CDK and network oscillators. *Nature* **453:** 944–947.

Sozzani R, Cui H, Moreno-Risueno MA, Busch W, Van Norman JM, Vernoux T, Brady SM, Dewitte W, Murray JA, Benfey PN. 2010. Spatiotemporal regulation of cell-cycle genes by SHORTROOT links patterning and growth. *Nature* **466:** 128–132.

Van Limbergen J, Wilson DC, Satsangi J. 2009. The genetics of Crohn's disease. *Annu Rev Genomics Hum Genet* **10:** 89–116.

Glossary

Readers should note that words in *italics* are themselves glossary terms and are defined herein.

actin microfilaments: A type of structural filament in the *cytoskeleton* of *eukaryotic* cells that is responsible for cellular movement and intracellular transport.

activation energy: The minimum amount of energy required for a chemical reaction to occur.

activator: A protein that binds to DNA to initiate production of a particular RNA molecule.

active site: A region within a folded, or mature, *enzyme* that binds to target molecules and enables enzyme activity.

adapters: Short sequences of DNA that are ligated to genomic fragments for various purposes, including for priming sequencing reactions.

additive: A type of interaction between two *genes* that causes the double mutant to exhibit a *phenotype* that is the sum of each individual gene's mutant phenotype.

adenosine diphosphate: An organic compound involved in energy transfer. ADP is a less-energetic adenine *nucleotide* than *ATP* and results from de*phosphorylation* of *ATP* by *enzymes* called ATPases.

adenosine triphosphate: An organic molecule that serves as the primary energy carrier in the cell.

ADP: See *adenosine diphosphate*.

Affymetrix: The company that pioneered direct synthesis of DNA probes onto a solid support for *microarray* production.

agamous (ag): A floral *gene* mutation causing *stamens* to become petals and *carpels* to become *sepals*.

alleles: Different forms (sequences) of the same *gene*.

α-helix: A helical form adopted by regions within a protein (see also *β-sheet*).

alternative splicing: A process occurring during *transcription* by which *exons* and *introns* are removed in variable ways, resulting in production of different *mRNA* species from a single *gene*.

Alu **retrotransposons:** A class of *transposable elements*, or DNA sequences capable of changing genomic position, that exist within the human *genome* and are associated with several inherited human diseases and forms of cancer.

amino acid: Organic monomers that are linked together during *translation* to make proteins.

amino-acyl site: During *translation* to make proteins, the amino-acyl site corresponds to the position within the *ribosome* where charged tRNA molecules pair with the correct *codon* in the messenger RNA.

anaphase-promoting complex: A complex of proteins that ligates the small regulatory protein ubiquitin to proteins to mark them for proteasome-mediated destruction.

annealing: The process by which two complementary, *single-stranded* nucleic acids come together by hydrogen bonding.

anticodon: The three-*nucleotide* sequence in a tRNA molecule that is complementary to the *codon* sequence, allowing the correct *amino acid* to be added to a growing protein.

APC: See *anaphase-promoting complex*.

apetala2 (*ap2*): A floral gene mutation causing sepals to become carpels and petals to become stamens.

apetala3 (*ap3*): A floral *gene* mutation causing *petals* to become *sepals* and *stamens* to become *carpels*.

apoptosis (or programmed cell death): The process by which cells die based on a genetically controlled intracellular program.

Arabidopsis thaliana: A species of plant that is commonly used to study *development* owing to its possession of a small, sequenced *genome* and a short generation time.

asymmetric division: A type of cell division that results in two daughter cells that go on to have different fates.

ATP: See *adenosine triphosphate*.

attractor: In a dynamic system, an attractor is the set toward which a variable evolves over time.

autocrine signaling: A mechanism of cellular signaling by which cells that emit a signal also perceive the signal (see also *paracrine signaling*).

balancing selection: Process of natural selection that results in the maintenance of more than one *allele* of a single *gene* within a population (see also *negative selection*).

basal transcription: The production of low levels of RNA that depends on binding of *general transcription factors* to regulatory DNA sequences (as opposed to highly regulated and specific transcription factors).

bases (adenine, thymine, cytosine, and guanine): Chemical components of DNA that, when bound to a sugar molecule, constitute *nucleotides*. Adenine, cytosine, and guanine are components of RNA as well, but thymine is replaced by *uracil* in RNA molecules.

basic local alignment search tool: An *in silico* tool that enables sequence alignment based on a heuristic algorithm that searches for small overlapping regions within a query sequence, assigns a score based on sequence similarity to a reference sequence, and matches high-scoring regions.

beam lines: High-energy X-ray beams targeted at protein crystals to generate a diffraction pattern that represents protein structure.

bell curve: A probability density function whose distribution is bell shaped.

β-sheet: A flat structure adopted by regions within a protein (see also *α-helix*).

biologicals: Pharmaceutical products made from living organisms, tissues, or cells.

biological replicate: Sample isolated from an independent, yet replicated, experiment in order to determine variation intrinsic to the system (see also *technical replicate*).

BLAST: See *basic local alignment search tool*.

body plan: Refers to the basic organization of organs and body parts within a developing organism.

Boolean: Basic network logic using "AND" or "OR" between nodes to describe regulatory relationships.

***Caenorhabditis elegans*:** A species of roundworm commonly used as a *model organism* in studies of animal *development*.

calmodulin (calcium-modulated protein): A messenger protein that activates target proteins when bound by calcium ions.

candidate gene: A *gene* hypothesized to control (or influence) the biological process under study.

capping: Chemical modification at the 5' end of an *mRNA* molecule that prevents degradation of the transcript.

carpels: Female sexual organ in flowers (see also *stamens*).

catalytic: Refers to the conversion of one substance to another by the activity of an *enzyme*.

cDNA: Copies of DNA ("complementary DNA") made from an mRNA template by means of *reverse transcriptase*.

CDK: See *cyclin-dependent-kinase*.

cell cycle: A series of steps that take place to prepare for and carry out cell division. Each of the four steps (see G_1, S, G_2, M) is called a "phase."

cellular state: The sum of all molecular interactions occurring in a cell at a given time.

centiMorgan (cM): A unit of measure that represents the distance between chromosomal positions and reflects genetic *linkage*. 1 cM is equivalent to a chromosomal region that has 0.01 crossovers on average per generation.

central dogma: The idea that, in all living organisms, DNA gives rise to RNA, which gives rise to protein, and that this flow of genetic information is unidirectional.

centromeres: Regions of *chromosomes* that attach to *microtubules* for separation during *mitosis* and *meiosis*.

chain-termination sequencing: A method of DNA sequencing involving termination of the copied sequence by incorporation of *dideoxynucleotides*.

chaperonins: Proteins that aid other proteins in acquiring the correctly folded state.

chemokine receptor: Membrane protein that binds small signaling proteins involved in immunity (chemokines).

ChIP-Seq: See *chromatin immunoprecipitation sequencing*.

chloroplast: A type of plant *organelle* that contains chlorophyll, a sunlight-absorbing molecule, and uses the energy of light to synthesize sugars.

chromatids: Newly copied and condensed chromosomes present during cell division.

chromatin: DNA and *histone* proteins assembled in a complex for the purpose of tight packaging within the nucleus.

chromatin immunoprecipitation sequencing: A method used to identify the sequences bound by DNA-binding proteins.

chromatin-remodeling complex: A protein complex that modifies *chromatin*.

chromatin-modifying enzymes: Proteins that catalyze the chemical modification of *histone* proteins for regulation of *gene expression* (e.g., methylation).

chromosomes: The DNA in a cell that contains the information to perform cellular processes. *Eukaryotic* DNA is not one continuous chain, but instead is organized into discrete pieces known as chromosomes.

***cis*-elements:** Regions of DNA that bind regulatory proteins to control *gene expression*.

***cis*-regulatory code:** Particular combinations of *transcription factors* required to bind to DNA in order to influence *gene expression*.

CNV: See *copy number variation*.

codons: Three-*nucleotide* sequences found in *mRNA* that code for specific *amino acids*.

complex traits: Traits that are controlled by more than a single *gene* (see also *monogenic traits*).

confocal microscopy: A type of microscopy that eliminates out-of-focus laser light with a dichromatic mirror and pinhole so that individual planes within a tissue can be visualized without the need for tissue-sectioning.

conformation: The physical orientation or shape of a protein.

contractile ring: A cellular structure consisting of *actin* and myosin that assists in division of cells.

copy number variation: Refers to regions of DNA that vary in copy number and are inherited within families.

cristae: Structures within the inner membrane of *mitochondria* where energy-producing chemical reactions occur.

cRNA: Copies of RNA made from *cDNA* templates.

Crohn's disease: A *quantitative trait* and intestinal disease that is controlled by several *QTLs*.

cyclin: A protein whose creation and destruction is tightly linked to progression through phases of the *cell cycle*.

cyclin-dependent kinase: A phosphorylating *enzyme* (*kinase*) whose activity depends on the presence of *cyclin* protein.

cytokines: Soluble signaling proteins that regulate diverse biological processes.

cytokinesis: Process in which one cell divides itself into two; the final step in the *cell cycle*.

cytoplasm: The aqueous substance within a cell.

cytoplasmic domain: The region of a cell that is enclosed by the cell membrane, but excluded by *organelles*.

cytoskeleton: Structural components of a cell, including *microtubules, actin* microfilaments, and *intermediate filaments*.

DARC: See *Duffy antigen/chemokine* receptor.

Dario rerio: A species of fish, known as zebrafish, commonly used as a *model organism* to study developmental and regenerative biology.

denticle belts: Hairlike structures that emerge as bands from the fruit fly embryo.

development: Collectively, all of the processes by which organisms grow.

dideoxynucleotides: Chemically modified *nucleotides* that terminate a growing strand of DNA owing to the absence of a 3'-hydroxyl group.

differentiation: Process of specialization of cells to take on certain fates and perform certain functions.

diploid: Refers to a *genome* with two copies of each *chromosome* in its unreplicated state (see also *haploid*).

disulfide bridge: A chemical bond formed between the sulfur atoms of two cysteine amino acid residues within a protein.

DNA (deoxyribonucleic acid): The molecule that contains all the genetic information for the growth, function and reproduction of an organism (see also *ribonucleic acid*).

DNA-binding domain: Region of a protein that binds directly to DNA.

DNA ligase: An *enzyme* that joins pieces of DNA.

DNA polymerase: An *enzyme* that adds complementary *nucleotides* during replication of a DNA template.

DNA primase: An *enzyme* that generates a short piece of complementary RNA to initiate replication of DNA.

DNA replication: The process by which cells copy DNA to ensure that all genetic information is transmitted to daughter cells.

double-stranded RNA: A sequence-specific tool used to disrupt *mRNA* expression.

Drosophila melanogaster: A species of fruit fly commonly used to study animal *development*.

dsRNA: See *double-stranded RNA*.

Duffy antigen/chemokine receptor: A *gene* that encodes a membrane-bound *receptor* that normally binds chemokine signaling molecules but also enables entry of the malarial parasite.

electron transport chain: A process that couples electron transfer between molecules to generation of an electrochemical proton gradient in order to generate energy.

electrospray ionization: A method used in mass spectrometry to ionize molecules.

elongation: The process of adding complementary *nucleotides* to the growing strand of DNA.

embryonic stem cells (ES cells): Cells of the embryo that are capable of self-renewal as well as *differentiation* into any tissue in the body (see also *stem cell*).

EMS: See *ethyl methanesulfonate*.

ENCyclopedia Of DNA Elements (ENCODE): A project aimed at identifying all regulatory regions in the human *genome*.

endocytosis: A process by which cells take in molecules that cannot pass through the cell membrane.

enhancers: Regions of DNA that bind regulatory proteins to increase expression of *genes*.

entrained: Refers to resetting of the circadian clock by altering light–dark cycles.

enzymes: Proteins that catalyze reactions or transform substances.

epigenetic: Modifications of *chromatin* that can be inherited, but do not change the genetic code.

epistatic: A type of interaction between two *genes* that is non*additive*; a double-mutant organism displays the *phenotype* of one of the single mutants (see also *additive*).

ESI: See *electrospray ionization*.

ethyl methanesulfonate: A chemical agent used to cause alkylation of guanine *bases* that results in *nucleotide*-substitution-mediated genetic mutation.

eukaryotic: Refers to organisms whose cells contain a nucleus.

even-skipped: Refers to a mutation that results in removal of even-numbered segments in a fruit fly embryo.

exit site: The site where a tRNA exits a *ribosome* following release of its *amino acid* cargo during protein synthesis.

exocytosis: The process by which cells expel molecules that cannot otherwise pass through the cell membrane.

exons: Regions of protein-coding sequence found in DNA (see also *introns*).

extracellular domain: Region of a membrane-bound protein that is found outside the membrane and frequently binds to signaling *ligands*.

feedforward loop: A *network motif* that represents interaction of two *transcription factors* with a downstream target such that activation of the first transcription factor results in activation of the second transcription factor, and together the first and second factors activate the target.

fermentation: A biochemical process that some organisms use to generate energy in the absence of oxygen.

folding: A process by which proteins take on a three-dimensional structure.

454 sequencing: A large-scale, pyrosequencing technology.

G_1 phase: The first phase of the *cell cycle* in which cells prepare to copy key molecules.

G_2 phase: The third phase of the *cell cycle* in which cells check for faithful replication of the *genome*.

gel electrophoresis: A method to separate molecules such as DNA or RNA according to their molecular weight.

gene: Refers to regions of DNA that include coding and regulatory sequences.

gene expression: The process by which a transcript is synthesized and/or protein is translated according to information encoded by the *gene*.

gene-regulatory network: A graphic representation of regulatory relationships between *transcription factors* and signaling molecules with their targets.

GeneChips: An *Affymetrix microarray* used to measure transcript expression.

general transcription factors: The *transcription factors* required to initiate *transcription* in *eukaryotic* organisms.

genetic map: A graphic representation of *genetic marker* positions relative to one another.

genetic markers: Differences in genetic sequence between individuals that "mark" *phenotypes* but do not necessarily influence them.

genetic screen: A method used to identify *genes* controlling altered *phenotypes*.

genetics: The study of how *gene* mutations affect *phenotypes*.

genome: All the DNA in all the *chromosomes* in an organism.

genome sequencing: The process of obtaining DNA sequence across the entire *genome* of an organism.

genome-wide association study: A method to identify regions of DNA that influence *complex traits*.

genome-wide expression analysis: Examination of the levels of expression of all genes at a given time or in a particular tissue.

glycolysis: Biochemical reaction carried out within cells to extract energy stored in carbohydrates.

G-protein complex: A *receptor* protein complex that assembles upon *ligand* binding, consisting of a *G-protein-coupled receptor* (GPCR) and three intracellular subunits (α, β, and γ). The intracellular subunits are found in both plants and animals, but GPCRs are found only in animals and fungi.

G-protein-coupled receptor: A type of membrane-bound *receptor* that transduces extracellular signals by means of formation of a *G-protein complex*.

GPCR: See *G-protein-coupled receptor*.

GRN: See *gene-regulatory network*.

growth factors: Molecules (usually proteins or *hormones*) that stimulate cellular growth, proliferation, and *differentiation*.

GTPase: An *enzyme* that removes a phosphate molecule from GTP.

GWAS: See *genome-wide association study*.

haploid: Refers to an organism, or gamete, that has one copy of each *chromosome* in its unreplicated state (see also *diploid*).

haplotypes: A region of DNA sequence that contains certain *alleles* that are inherited together owing to a lack of *recombination*.

HapMap Project: An effort to identify common genetic variations within human populations.

hidden Markov models: A statistical model that assumes a Markov process with unobserved states.

high-pressure liquid chromatography: A method to separate compounds into classes (e.g., *amino acids*, sugars, or lipids).

His tag: The addition of several histidine *amino acids* to a *recombinant* protein to allow for protein isolation and purification.

histones: Proteins found in *eukaryotic* cells that bind to and organize DNA into *nucleosomes*.

HMMs: See *hidden Markov models*.

homeobox: A class of *transcription factors* that control *gene expression* in early stages of *development*.

homologous recombination: The process whereby sister *chromatids* "cross over" one another to exchange regions of DNA.

hormones: Small signaling molecules that can travel long distances in the body to control cellular processes.

HPLC: See *high-pressure liquid chromatography*.

Human Epigenome Project: A project aimed at mapping all *chromatin* modifications in the human *genome*.

Human Genome Project: A project aimed at mapping all *genes* in the human *genome*.

Illumina sequencing: A next-generation sequencing technology that uses amplification of DNA that has been fixed to slides.

indeterminate: Refers to growth that is continuous.

interferon: A *cytokine* signaling molecule involved in immunity.

intermediate filaments: A component of the *cytoskeleton* that provides structural support to the cell.

introns: Regions of DNA that do not code for protein, but often serve as *gene-regulatory* sequences (see also *exons*).

JAK: See *Janus kinase*.

Janus kinase: An intracellular *kinase* that transduces *cytokine* signals.

kinase: An *enzyme* that regulates target proteins by adding a phosphate molecule.

Krebs cycle: A set of chemical reactions that generate ATP and reducing agents.

lactase-phlorizin hydrolase: An *enzyme* that breaks down lactose and allows adults to digest milk products after weaning.

lagging strand: Refers to the nascent DNA strand that is positioned with the 3' end in the open DNA helix and is replicated discontinuously.

leading strand: Refers to the nascent DNA strand that is positioned with the 5' end in the open DNA helix and is replicated continuously.

library: A collection of DNA fragments used for various purposes, including sequencing.

ligand: A molecule that is usually a small protein or *hormone*, which binds to a *receptor* in order to transduce a signal.

linkage: The concept that *genes* in relatively close proximity to one another on the *chromosome* will be inherited together.

lipid bilayer: Lipids have a long hydrophobic domain and a small charged "head" region that, in an aqueous environment, will spontaneously aggregate such that the hydrophobic regions remain together and the charged heads lie apart—two-layered membranes like this constitute most cell membranes.

M phase: See *mitosis*.

machine learning algorithms: Computational tools that find patterns based on a training set (e.g., images or DNA sequences).

major groove: Structural regions throughout the DNA helix that are large enough for *transcription factors* to access and bind to *nucleotides*.

MALDI: See *matrix-assisted laser desorption/ionization*.

MAP kinases: See *mitogen-activated protein kinases*.

map-based strategy: An approach used for whole-*genome sequencing* that requires overlapping fragments of sequence.

mass spectrometry: A method to determine the mass to charge ratio of ions.

matrix-assisted laser desorption/ionization: A type of *mass spectrometry* that uses laser ionization to vaporize molecules.

mediator: A protein that enables association between *general transcription factors* and those factors bound to *enhancers* to cause increased *gene expression*.

meiosis: A cellular process similar to *mitosis*, but resulting in *haploid* daughter cells for the purpose of gamete production for sexual reproduction (see also *M phase*).

melanocortin 1 receptor gene (melanocyte-stimulating hormone receptor, *MC1R*): Encodes a membrane-bound protein that binds to signaling *ligands* on the surface of the melanocyte cell and is responsible for some variability in human pigmentation.

melting: The process of separating double-stranded DNA into single strands.

Mendelian: Refers to traits that are controlled by a single *gene*.

messenger RNA (mRNA): The type of *RNA* molecule that encodes proteins and is referred to as a "transcript."

metabolic network: A graphical representation of the relationships between *enzymes* and metabolites.

metabolomics: The study of all metabolites in a system; a method to identify all metabolites in a system.

microarray: A slide fixed with thousands of *single-stranded DNA oligonucleotide* "probes" used to understand how/when *genes* are expressed.

microRNA (miRNA): A class of small, noncoding RNA molecules that can regulate *mRNA* stability and translation in a sequence-specific manner.

microtubules: A type of cytoskeletal protein polymer involved in many cellular processes, including cell motility and intracellular transport.

mitochondria: *Organelles* found in *eukaryotic* cells whose main role is to produce energy in the form of *ATP*.

mitogen-activated protein kinases: A class of serine/threonine *kinases* that control many cellular processes, including *gene expression*.

mitosis: The phase in the *cell cycle* where sets of replicated chromosomal DNA are aligned and separated into two daughter nuclei just as *cytokinesis* commences (see also *meiosis*).

mitotic cyclin: The *cyclin* protein whose presence/absence controls entry into *mitosis*.

model organisms: Species chosen to represent organisms within their kingdom of life that are used to study biological processes (e.g., *Arabidopsis*, mouse).

monogenic traits: Traits controlled by a single *gene* (see also *complex traits*).

morphogen gradient: Refers to factors that control aspects of *development* and exist in gradually increasing or decreasing concentrations across a tissue or throughout the organism (usually in the embryo).

morpholino antisense oligonucleotides: An artificial chain of stable, *single-stranded* DNA that is used to reduce *mRNA* stability and protein production, in a sequence-specific manner.

multiplexing: An approach used to combine multiple DNA samples in the same sequencing run through the addition of sample-specific barcode *adapters*.

multiple reaction monitoring: A method for analyzing proteins with mass spectrometry that allows accurate measurement of protein concentrations.

***Mus musculus*:** A species of mouse commonly used for studies in animal developmental biology.

NADP: See *nicotinamide adenine dinucleotide phosphate*.

NADPH: The reduced form of *NADP* produced at the end of the *electron transport chain*, which can be used in the Calvin cycle to produce carbohydrates.

nanopore: A very small physical channel that can be used to detect differences in the *bases* of DNA as the DNA passes through the channel, and thus a potential means of determining *nucleotide* sequences.

negative selection: In natural selection, this refers to the removal of deleterious *alleles* from populations (see also *balancing selection*).

network motifs: Graphical representations of regulatory relationships, usually between a limited number of *transcription factors* and target *genes*.

neutral selective pressure: In natural selection, regions that are noncoding and do not influence expression of *genes* accumulate mutations at a background rate and are not selected for or against.

nicotinamide adenine dinucleotide phosphate: *NADP* is used in many cellular reactions as it can be reduced to *NADPH*, a key source of reducing power in the Calvin cycle.

NMR: See *nuclear magnetic resonance*.

normal distribution: A probability density function that is bell shaped.

northern blotting: A method used for detection of RNA molecules that depends on *gel electrophoresis*, followed by transfer of RNA to a membrane and visualization of specific RNA species by using labeled probes.

nuclear magnetic resonance: A method used to understand protein structure that is based on spectra emitted from atomic nuclei within proteins following excitation by electromagnetic energy.

nuclear pores: Channels in the nuclear membrane created by assembly of pore-forming protein complexes.

nucleolus: A region within the nucleus that contains large amounts of DNA that specifically encodes ribosomal RNA.

nucleoplasm: Aqueous substance within a cell's nucleus where most DNA is found.

nucleosomes: *Histone*-protein complexes that bind to DNA and, collectively, constitute chromatin.

nucleotide: A molecule that consists of a *base*, a sugar, and at least one phosphate group—the individual components in DNA and RNA.

odd-skipped: Refers to a mutation that results in removal of odd-numbered segments in a fruit fly embryo.

Okazaki fragments: Short segments of newly synthesized DNA on the *lagging strand*.

oligonucleotide: A polymer consisting of more than one *nucleotide* (but usually fewer than 100 nucleotides).

oncogene: A *gene* that causes cancer when activated (see also *tumor suppressor*).

1000 Genomes Project: A project aimed at obtaining whole-*genome* human sequence data from 1000 individuals from across the world using next-generation sequencing technology.

open reading frames: Regions of DNA that code for protein.

ORFs: See *open reading frames*.

organelles: Compartments within *eukaryotic* cells that have specific functions.

p21: A protein that can be activated to halt *cell cycle* progression.

p53: A *tumor-suppressor* protein that receives multiple inputs and can alter *gene expression* to halt *cell cycle* progression following DNA damage or initiate *apoptosis*.

paracrine signaling: A mechanism of cellular signaling by which cells perceive an extracellular signal (see also *autocrine signaling*).

PCR: See *polymerase chain reaction*.

peptide bond: A covalent bond between *amino acids* in a protein formed in the *ribosome* during *translation*.

peptidyl site: During *translation*, the peptidyl site corresponds to the position within the *ribosome* where the *peptide bond* is formed between a growing protein and a new *amino acid* (see also *exit site*).

permissive temperature: The temperature at which a *temperature-sensitive mutant* grows normally (see also *restrictive temperature*).

petals: Structures of a flower that surround the reproductive organs (see also *sepals*).

phenotypes: Observable changes in mutant organisms when compared with wild-type (or normal) counterparts.

phosphate backbones: String of molecules of phosphate that together form the outer edges of the DNA helix.

phosphorylation: The addition of a phosphate group to a protein or other molecule, usually catalyzed by a *kinase*.

photolithography: A method used to synthesize *oligonucleotide* probes directly onto a solid support.

photosynthesis: A process performed in *chloroplasts* of some cells that converts light to chemical energy.

photosystem II: A protein complex housed in the *chloroplast* that loses an electron to the *electron transport chain* when chlorophyll absorbs sunlight.

pistillata (*pi*): A floral *gene* mutation causing *petals* to become *sepals* and *stamens* to become *carpels*.

plasmodesmata: Pores in the plant cell membrane that allow transport of molecules between cells.

poly(A) tails: A series of adenosine *nucleotides* added to the 3′ end of nascent *mRNAs* to prevent degradation.

polymerase chain reaction: A technique used to amplify, or copy in an exponential manner, double-stranded DNA *in vitro*.

population structure: The concept that populations are genetically subdivided in some ways (e.g., the frequency of *alleles* present can vary within and between populations).

positive selectable marker: A tool used in transgenics as a means of ensuring expression of the introduced DNA.

PPi: See *pyrophosphate*.

prebranch sites: Specified sites on a growing plant root from where a lateral root can be formed.

primary metabolites: Metabolites that are directly involved in normal growth and *development*.

primary mRNA: A transcript that includes *introns* as well as *exons*.

primary structure: The sequence of *amino acids* in a protein.

primer: A short piece of DNA or RNA that anneals to its complementary template sequence, providing a short double-stranded region adequate for polymerase binding (see also *sequencing primer*).

production sequencing: The second step in genomic sequencing, which includes generating small DNA fragments, running the sequencing reaction and analyzing it on an automated sequencer.

proenkephalin-B (prodynorphin) (*PDYN*): A *gene* expressed in the brain that has been associated with schizophrenia and temporal lobe epilepsy.

prokaryotic: Pertaining to the prokaryote kingdom of life, which includes all organisms (bacteria) that lack membrane-bound *organelles* (see also *eukaryotic*).

promoter: DNA region immediately upstream of *transcription* start sites whose sequence, in part, directs binding of *transcription factors* and expression of adjacent *genes*.

protease: An enzyme that degrades protein.

pyrophosphate: A molecule consisting of two phosphate groups that is released upon *nucleotide* addition to a growing strand of DNA or RNA.

pyrosequencing: A method of DNA sequencing that detects *pyrophosphate* release.

QTLs: See *quantitative trait loci*.

quantitative trait: A trait that can be quantified that is continuously distributed across a population (e.g., seed size).

quantitative trait loci: Chromosomal regions that are associated with *quantitative traits*.

quaternary structure: The complex of proteins required to carry out a particular function.

Ras: A *GTPase enzyme* that, when mutated, can cause uncontrolled growth, as in some cancers.

Ras pathway: A *signal transduction pathway* that perceives an extracellular signal, auto-phosphorylates to recruit *Ras*, and eventually alters *gene expression* to enhance growth.

rational network design: Construction of biological networks to perform a desired function in cells.

receptors: Proteins that receive and transduce signals delivered by *ligand* binding.

recessive: Refers to a genetic mutation that requires two mutant copies of the *gene* to display the mutant *phenotype* in a *diploid* organism.

recombinant DNA technology: Methods to synthesize and introduce DNA from one organism into another.

recombination: A process whereby *chromosomes* exchange regions of DNA.

repetitive DNA elements: Segments of DNA that have similar sequences and are found throughout the *genome*.

replicates: Refers to experiments that are repeated to gauge noise inherent to the system (see also *biological replicate* and *technical replicate*).

replication fork: The region of DNA that is opened or "unzipped" to allow *DNA replication* to occur.

repressilator: A synthetic network involving three short-lived, transcriptional *repressors*, whose output results in oscillatory behavior.

repressor: A protein that represses target function (e.g., a transcriptional repressor will repress target *gene expression*).

restrictive temperature: The temperature at which a *temperature-sensitive mutant* exhibits a *phenotype* (see also *permissive temperature*).

retrotransposons: Regions of DNA that "move" by transcribing an RNA copy and inserting themselves throughout the *genome*.

reverse genetics: An approach to study how perturbation of specific *genes* affects *phenotype*.

reverse transcriptase: An *enzyme* that builds complementary DNA (*cDNA*) from an RNA sequence.

ribonucleic acid (RNA): A molecule composed of *nucleotides*, but using *uracil* in place of thymine and is single stranded—*mRNA* is used to transfer information from linear DNA into three-dimensional proteins.

ribosomes: Large molecular complexes composed of proteins and ribosomal RNA that catalyze protein synthesis, or *translation*.

RNA interference (RNAi): A technique to introduce *dsRNA* into an organism to reduce expression of specific *mRNA* sequences.

RNA polymerase: An *enzyme* that catalyzes production of RNA from a DNA template.

RNA secondary structure: Because RNA molecules are *single-stranded*, they often readily fold-back on themselves and base pair to form secondary structure that can aid in stability and can be necessary for function (e.g., tRNA) (see also *secondary structure*).

RNase H: An *enzyme* that degrades an RNA *primer* so that it can be replaced with DNA during replication.

RNA-Seq: A method to obtain RNA expression information across the entire *genome* using next-generation sequencing.

RuBisCo: A highly abundant *enzyme* in plants that catalyzes the conversion of carbon dioxide into sugar.

S-phase: The phase in the *cell cycle* when DNA and other molecules are copied.

second messengers: Small molecules that transmit signals within the cell.

secondary structure: Refers to the presence of, for example, *α-helix* and *β-sheet* conformations within a protein.

secretion: An energy-requiring process that cells use to release molecules that do not readily pass through the cell membrane.

selected ion monitoring: A method used in mass spectrometry to reduce the complexity of the compounds being analyzed (see *multiple reaction monitoring*).

selective sweep: In natural selection, a *gene* and neighboring sequence will exhibit little sequence variation across a population if a mutation in that *gene* confers a fitness advantage.

semiconservative: Refers to *DNA replication* as one strand is newly synthesized ("nascent") and one strand is "conserved".

sepals: Leaf-like structures that surround the *petals* of a flower.

sequence depth: In sequencing, the number of *sequence reads* per region of DNA or RNA.

sequence reads: In sequencing, the sequenced copies of DNA or RNA regions.

sequencing primer: A short piece of DNA that is complementary to a template sequence and allows for binding of *DNA polymerase* and initiation of the sequencing reaction.

short interfering RNA (siRNA): Small RNA molecules that can regulate the stability or degradation of *mRNA* in a sequence-specific manner.

signal transduction pathway: A series of signaling events beginning with perception of an extracellular signal and leading to modulation of a cellular process.

signaling cross talk: Intersection between signaling pathways often attributable to shared network nodes.

single-molecule sequencing technologies: A type of next-generation sequencing technology that promises longer read-lengths and might dramatically decrease sequencing time.

single-nucleotide polymorphism: Variation in one *nucleotide* position between individuals that is often used as a *genetic marker*.

single-stranded: Refers to single chains of *nucleotides* that are not annealed to complementary sequences.

SNP: See *single-nucleotide polymorphism*.

somatic cells: Any cell in an organism that is not a germ cell.

spliced mRNA: *mRNA* that does not have *introns*.

splicing: A process to remove *introns* from *primary mRNAs*.

stamens: The male, or pollen-producing, reproductive organs of a flower (see also *carpels*).

STAT pathway: A signaling pathway that begins *signal transduction* upon *ligand* binding to the STAT cell-surface *receptor* and its subsequent dimerization.

stem cells: Cells that are capable of self-renewal and can divide to give rise to many cell types (see also *embryonic stem cells*).

***Strongylocentrotus purpuratus*:** A species of sea urchin commonly used to study early events in *development*.

structural genomics: A field of study aimed at understanding the three-dimensional structure of all proteins encoded in the *genome*.

synthetic biology: A field of study focused on design of synthetic networks to understand how networks function normally and to engineer new processes.

synthetic cell: A cell whose *genome* is entirely synthetic but can grow and divide.

tag: Refers to a short stretch of *amino acids* conjugated to a protein of interest, for the purposes of protein visualization or isolation with an antibody that recognizes the tag.

TATA box: A short region of DNA usually located 35 *nucleotides* upstream of the *transcription* start site that is recognized and bound by *TFIID* (a *general transcription factor*).

technical replicate: A single sample replicated within an experiment in order to determine noise introduced by the technique (see also *biological replicate*).

temperature-sensitive mutant: A mutant organism whose *phenotype* is not different from that of the wild type unless placed at a particular temperature (*restrictive temperature*).

template strand: The conserved strand of DNA from which a complement is built by *DNA polymerase* during replication or a *primary mRNA* is made during transcription.

tertiary structure: The three-dimensional structure of a folded protein.

TFIID: A *general transcription factor* that binds to the *TATA box* within the *promoter* of a *gene*.

thylakoids: Structures found within *chloroplasts* that house chlorophyll.

time of flight: A method to identify a compound's constituent components based on the time it takes for charged molecules to "fly" from an ionizer to a detector.

TOF: See *time of flight*.

transcription: A cellular process whereby template DNA sequence is used to produce *mRNA* "transcripts."

transcription factors: Proteins that influence *gene expression* by binding DNA either directly or indirectly.

transfer RNA (tRNA): A type of noncoding RNA that translates *mRNA* sequence into protein by carrying a specific *amino acid* cargo to the *ribosome* and binding based on a specific *anticodon* sequence.

transformation: A method used to introduce foreign DNA into a cell.

translation: A cellular process whereby template *mRNA* sequence is used to direct protein production in the *ribosome*.

transmembrane domain: A hydrophobic region within a protein that is embedded in the cell membrane.

transposable elements/transposons: Pieces of DNA that have the ability to move to other regions in the *genome* by expressing *transposase*.

transposase: An *enzyme* capable of excising sections of DNA and reinserting them elsewhere in the *genome*.

tumor suppressor: A *gene* that can cause cancer when its activity is reduced (e.g., when the *gene* itself is mutated or an upstream *activator* of the tumor suppressor is mutated) (see also *oncogene*).

tyrosine kinase receptor: A membrane-bound protein capable of transducing extracellular signals by means of *phosphorylation* of tyrosine residues.

uracil (U): A *base* used in place of thymine by *RNA polymerase* in the production of RNA.

vesicles: Small, membrane-bound "packages" that can be used to move cargo around the cell in an environment (pH, etc.) that can differ from that of the surrounding *cytoplasm*.

whole-genome shotgun: A method of whole-*genome sequencing* based on sequencing small fragments of DNA that need to be assembled into a contiguous sequence *post hoc*.

whorls: An arrangement of floral structures that radiate from a central point.

wild type: A normal organism against which mutants can be compared.

X-ray crystallography: A method used to determine protein structure based on X-ray diffraction of a crystallized protein.

yeast one-hybrid method: A method used to identify *transcription factors* that bind to a DNA sequence of interest.

yeast two-hybrid method: A method used to identify proteins that bind to one another.

Illustration Credits

Chapter 2: **Figs. 2.1–2.6, 2.8–2.10**, Modified, with permission, from Pearson Education, from Watson JD, et al. 2008. *Molecular biology of the gene,* 6th ed. (Figs. 5-1, 2-5, 6-1, 2-8, 8-2, 2-6, 8-11, Table 2-3, Fig. 7-3). **Figs. 2.7 and 2.11**, Modified, with permission, from Pearson Education, from Hardin J, et al. 2012. *Becker's world of the cell,* 8th ed. (Figs. 19-11, 18-12). **Figs. 2.12–2.14**, Modified, with permission, from Benfey PN, Protopapas AD. 2005. *Genomics* (p. 55, p. 57, p. 89).

   Chapter 3: **Figs. 3.1–3.5, 3.7**, Modified, with permission, from Pearson Education, from Watson JD, et al. 2008. *Molecular biology of the gene,* 6th ed. (Figs. 6-29, 12-1, 16-1, 17-9, 17-2, 17-11b). **Fig. 3.6**, (*A*) Modified, with permission, from Pearson Education, from Watson JD, et al. 2008. *Molecular biology of the gene,* 6th ed. (Fig. 7-18a); (*B*) Luger K, Mäder AW, Richmond RK, Sargent DF, Richmond TJ. 1997. *Nature* **389:** 251–260 (Fig. 7-18b). Image prepared with MolScript, BobScript, and Raster 3D by Leemor Joshua-Tor, used with permission. **Figs. 3.8 and 3.9**, Modified, with permission, from Pearson Education, from Benfey PN, Protopapas AD. 2005. *Genomics* (p. 43, p. 137). **Fig 3.10**, Adapted from Brown PO, Botstein D. 1999. *Nat Genet* **21:** 33–37; originally from Spellman PT, et al. 1998. *Mol Biol Cell* **9:** 3273–3297, with permission from the American Society for Cell Biology. **Fig 3.11**, http://en.wikipedia.org/wiki/RNA-Seq.

   Chapter 4: **Figs. 4.1, 4.3B, 4.4–4.5, 4.8–4.9, 4.12**, Adapted, with permission, from Pearson Education, from Hardin J, et al. 2012. *Becker's world of the cell,* 8th ed. (Figs. 4-5, 4-9, 4-11, 4-14, 9-6, 9-7, Table 15-1). **Fig 4.6**, Redrawn Shutterstock image #169460597. **Fig. 4.11**, Adapted from http://prodomweb.univ-lyon1.fr/priam/REL_JUL03/MELILOTI_ALL/PRIAM_REPLICON/RES_PRIAM/map00290_org.html, courtesy the PRIAM database, http://priam.prabi.fr, maintained by the Pôle Rhône-Alpin de Bioinformatique; Claudel-Renard C., et al. 2003. *Nucleic Acids Res* **31:** 6633–6639. **Fig. 4.13**, Modified, with permission, from Michael W. Davidson, National High Magnetic Field Laboratory, Florida State University http://www.olympusmicro.com/primer/techniques/confocal/confocalintro.html. **Fig. 4.14**, Reproduced from Perlman ZE, et al. 2004. *Science* **306:** 1194–1198, with permission from AAAS. **Fig. 4.15**, Modified, with permission, from Pearson Education, from Benfey PN, Protopapas AD. 2005. *Genomics* (p. 393, p. 395 *top*).

   Chapter 5: **Figs. 5.1–5.2, 5.7–5.8**, Modified, with permission, from Pearson Education, from Watson JD, et al. 2008, *Molecular biology of the gene,* 6th ed. (Figs. 14-14, 14-4 *left*, 14-7, 15-37 *far right*, 17-22a, 17-22b). **Figs. 5.3, 5.6, 5.9, 5.10**, Modified, with permission, from Pearson Education, from Hardin J, et al. 2012, *Becker's world of the cell*, 8th ed. (Figs. 22-2, 6-7, 14-16, 14-5). **Fig 5.4**, Source unknown. **Figs. 5.5, 5.11–5.14**, Modified, with permission, from Pearson Education, from Benfey PN, Protopapas AD. 2005. *Genomics* (p. 267 *lower middle*, p. 391, p. 239 *lower middle*, p. 249, p. 257).

   Chapter 6: **Figs. 6.2, 6.4**, Courtesy of Steve Haase, Duke University. **Figs. 6.3, 6.5**, Adapted from Morgan DO. 2007. *The cell cycle: Principles of control,* © New Science Press Ltd. in association with Oxford University Press and Sinauer Associates, Inc. (p. 51, p. 55), with permission from Oxford University Press. **Fig. 6.6**, Data from Orlando DA, et al. 2008. *Nature* **453**: 944–947. **Fig. 6.7**, Reproduced from Orlando DA, et al. 2008. *Nature* **453**: 944–947, with permission from Macmillan Publishers Ltd. **Fig. 6.9**, Reproduced from Moreno-Risueno MA, et al. 2010. *Science* **329**: 1306–1311. **Fig. 6.10**, Reproduced from Moreno-Risueno MA, et al. 2010. *Science* **329**: 1306–1311.

   Chapter 7: **Fig. 7.1**, Modified, with permission, from Pearson Education, from Benfey PN, Protopapas AD. 2005. *Genomics* (p. 383 *middle*). **Fig. 7.2**, Courtesy Benfey laboratory.

   Chapter 8: **Figs. 8.1–8.4, 8.6–8.9**, Courtesy of Elliot Meyerowitz, California Institute of Technology, Pasadena. **Fig 8.3**, Reprinted from Meyerowitz EM, et al. 1991. *Dev Suppl* **1**: 157–167, with permission from Company of Biologists. **Fig. 8.4**, Reprinted from Bowman JL, et al. 1989. *Plant Cell* **1**: 37–52, with permission from the American Society of Plant Biologists.

   Chapter 9: **Fig. 9.1**, Courtesy of Courtney Babbitt from the UCSC Genome Browser. **Table 9.1**, Adapted from Haygood R, et al. 2007. *Nat Genet* **39**: 1140–1144, with permission from Macmillan Publishers Ltd.

Index

Page references followed by f denote figures; those followed by t denote tables.

A

ABC model of flower development, 112–115
Actin microfilaments, 58f, 59, 94, 96, 133
Activation energy, 72, 133
Activator, 31, 31f–32f, 33, 85, 106, 133
Active site, 72, 73f, 133
Adapters, 21, 133
Additive genetic interaction, 115, 133
Adenine, 4, 4f–5f
Adenosine, in poly(A) tails, 29
Adenosine diphosphate (ADP), 52–53, 52f, 54f–55f, 133
Adenosine triphosphate (ATP)
  defined, 133
  in glycolysis, 53–54, 54f–55f
  synthesis in Krebs cycle, 54, 56f
  synthesis in photosynthesis, 52, 52f
  use in Calvin cycle, 52, 53f
Affymetrix, 38, 133
Agamous (ag) mutation, 111–114, 111f, 112t, 113f–115f, 133
Agarose, 15
Alleles, 14, 133
Alon, Uri, 86
α-helix, 70, 134
Alternative splicing, 14, 41–42, 41f, 134
Alu retrotransposons, 124, 134
Amino acids

defined, 134
genetic code and, 12–13, 12f
peptide bond between, 70, 143
protein structure and, 70–71, 71f–72f
role in translation, 68–69, 68f–69f
Amino-acyl site, 69, 69f, 134
Anaphase-promoting complex (APC), 92–93, 93f, 95f, 134
Annealing, 8–9, 19, 20f, 134
Antibodies, 84
Anticodon, 68–69, 69f, 134
Antisense oligonucleotides, morpholino, 106
APC (anaphase-promoting complex), 92–93, 93f, 95f, 134
Apes, humans compared to, 121
Apetala2 (ap2) mutation, 111–114, 111f, 112t, 113f–115f, 134
Apetala3 (ap3) mutation, 111–114, 112–115f, 112t, 134
Apoptosis (programmed cell death), 79–80, 79f, 134
Arabidopsis thaliana
  described, 134
  flower organization, 110–114, 110f–115f, 112t
  as model organism, 104
  oscillating expression of genes in roots, 101f–102f
  root development in, 107–108, 108f
Asymmetric division, 95, 134
ATP synthase, 52f